DATE DUE

The Gene Doctors

The GENE DOCTORS

Medical Genetics at the Frontier

YVONNE BASKIN

WILLIAM MORROW AND COMPANY, INC.
NEW YORK   1984

To my parents,
Lee and Leroy Baskin,
with love and gratitude for good genes
and a loving home

Library of Congress Catalog Card Number: 83-61743

ISBN: 0-688-02645-1

Printed in the United States of America

First Edition

1 2 3 4 5 6 7 8 9 10
BOOK DESIGN BY ELLEN LO GIUDICE

Acknowledgments

I am indebted to many people for information, assistance and advice that helped to shape this book.

Special thanks go to Theodore Friedmann, who introduced me to the potential of gene therapy, guided my first efforts to understand the science behind it, and reviewed several chapters of the manuscript.

I am also grateful to Diane Lindquist and Paul Saltman, who both read the entire book in manuscript and offered valuable criticisms.

Many geneticists and molecular biologists generously granted interviews during the past three years and took the time to discuss their work in progress. This has made it possible for me to provide up-to-date information on a science that is changing daily and has not yet been canonized in textbooks. Among them are Richard Axel, Richard D. Palmiter, Richard Mulligan, Thomas Maniatis, Jon W. Gordon, Phillip A. Sharp, Stanfield Rogers, Martin Cline, W. French Anderson, Leroy Hood, Victor Mckusick, Raymond L. White, Arlene Wyman, and Mary E. Harper.

I acknowledge with gratitude the suggestions and reactions provided by scientists who reviewed chapters: John O'Brien, Richard D. Palmiter, Richard Axel, Richard Mulligan, Mary E. Harper, Raymond L. White, Stanfield Rogers, and Phillip A. Sharp. Obviously no failures of mine that remain should be counted against any of them.

This book also owes much to the hospitality extended by Aileen Smart and Tom Stewart during the research; to Peter Lambrou for encouragement that has special significance coming from a fellow author wrestling with his own book; to David McHam, who first taught me what it takes to make a life of writing; and especially to Jon for his daily support and understanding throughout. To all of these, my deepest thanks.

Contents

I
GENETIC
CHOICES

1. Gene Therapy: Expanding Our Options

Man can create in his imagination worlds different from the actual one and can visualize himself in these imaginary worlds . . . The adaptive value of forethought or foresight is too evident to need demonstration. It has raised man to the status of the lord of creation.

Self-awareness and foresight brought, however, the awesome gifts of freedom and responsibility . . . This is a dreadfully heavy load to carry. No other animal has to withstand anything like it.

—THEODOSIUS DOBZHANSKY, 1962

The hum of ventilating fans from two large, dial-covered machines soothes the dimly lit room of the clinic. A woman, her exposed belly visibly beginning to swell, lies stretched out on a narrow bed beside the machines, alert and a little nervous. An ultrasound technician draws a microphone-like wand back and forth across the woman's lower belly, probing her womb with inaudible sound waves the way sonar scans the deep oceans. On a TV screen built into one of the machines, the woman and the technician watch as the reflected sound waves begin to generate a fuzzy image of a four-month-old fetus. The technician spots a pocket of amniotic fluid safely to the side of the tiny, five-inchlong form, and X's the spot on the woman's stomach.

Then the lights flick on, a doctor gloves up, and the outside world prepares to make contact with the new life for the first time. The probing takes less than a minute. Through abdominal skin numbed by anaesthetic, the physician punches a hollow fourinch-long needle, pushing it all the way into the womb, attaches a syringe, and draws up two tablespoons of the pale yellow fluid surrounding the fetus. Within the fluid are cells, enzymes, and other substances shed by the growing child—evidence the geneticist needs to determine what sort of stuff this new life is made of.

The fetal cells, separated out, nourished, and incubated in a lab dish, will be allowed to grow and multiply until the scientists have enough to work with. But it's not the cells themselves that tell the story. Within the nucleus of each cell is an identical load of DNA containing the inherited instructions from which the new life is being assembled. The DNA blueprint, made up of units called genes, obviously specifies that this new life will be a male or female human being, not a lobster or a geranium. But it's the finer details of the instructions that interest the geneticist, and the prospective parents, during this fetal screening: Is there a glitch (mutation) in the instructions somewhere that could foul up any of the hundreds of biochemical processes that are needed to carry on a normal human life? A glitch that later in life will cause toxic wastes to go unprocessed, invading germs to enter unchecked, or red blood cells to collapse prematurely? Will the glitch retard the baby's developing mind, clog its lungs with mucus, prevent its blood from clotting, stunt its growth, or waste its muscles? Will it leave this new individual especially vulnerable to allergies, heart attack, lung cancer, mental depression, arthritis, or diabetes?

In less than a month, the answers are in. A medical geneticist meets with the pregnant woman and her husband:

"First, since you asked, I can tell you you're going to have a girl. I can see you're not at all disappointed! As for the test results, let me make it clear that our tests can't detect every possible abnormality. Still, we've screened your baby's genes for several hundred of the most common defects or trouble spots that we know about. Automation has made that kind of broad screening practical, even though we have no reason to believe either of you is a carrier for most of the defects we checked for.

"The most serious trouble spot we found is the one you both were worried about—the risk of premature heart attacks. I'm afraid your daughter has inherited a gene for that trait from each of you. Now, since both of you are merely carriers—you each have one 'good' gene that mutes some of the influence of the 'bad' gene—you've been able to keep your risk of heart attacks in the normal range just by watching your life-style and taking medication. But your baby will have two of the abnormal genes and she could face a heart attack as early as childhood.

"Fortunately, that's a gene we've had good success in transplanting. Several of the large pharmaceutical firms supply copies of the gene already linked to a harmless virus. We can wait a few months after your baby is born, then remove a small amount of her bone marrow, infect it, and return it to her. The procedure is fairly simple and effective, but of course it's not cheap. At this stage you still have the choice of terminating this pregnancy and trying again.

"I'm glad we have some options to offer you in this case, but the choice is still up to you."

The encounter is fictional—but not far from reality. Ultrasound scanning and amniocentesis—the withdrawal of fetal cells from the womb for genetic screening—are already offered routinely in major medical centers around the world. Nearly two hundred hereditary diseases and another one hundred chromosome defects can be detected before birth today, and the number is growing rapidly. Automated screening systems are on the drawing board. The one thing physicians cannot do yet, however, is to fix routinely what they find. Aborting and trying again—or bearing a child with a serious, perhaps fatal handicap—are the only options available today to a woman who learns she is carrying a child with an inherited defect.

No one is satisfied with those options, and we may not be limited to them much longer. Advances in genetic engineering have brought us to the threshold of a new era in both the detection and treatment of hereditary disease. As our capacity to screen fetuses and identify defects before birth expands, the pressure for effective treatments at the genetic level is growing stronger. Medical researchers have already made several attempts to transplant new genes into human patients or to alter the activity of the patients' own genes. Gene therapy is medicine's next frontier.

2. The Frontier Within

It does not seem unrealistic to expect that, as more is
learned about control of cell machinery and heredity, we
will see the complete conquest of many of man's ills,
including hereditary defects in metabolism and the
currently more obscure conditions such as cancer and the
degenerative diseases, just as diseases of bacterial and viral
etiology are now being conquered.
  —EDWARD L. TATUM, 1958 Nobel Prize lecture

Not until the twentieth century did medicine finally get down to
causes, striking at the roots of disease. It was a revolution. For
thousands of years our forebears, not knowing the causes of their
distress, had lived in fear of the night air, strange miasmas, super-
natural powers, and vague disruptions in the body's humors.
Armed with an often frightening array of herbal and mineral elix-
irs, poultices and plasters, leeches, sweats, purgatives, and ca-
thartics, physicians could minister only to the symptoms of
disease—pain, fever, nausea, weakness, growths, and eruptions.
When patients with serious ailments did survive, it was often in
spite of the medical care they received.

The breakthrough came in the mid-nineteenth century with the
discovery by French chemist Louis Pasteur that germs cause infectious diseases. For the first time, blame for the plagues and
epidemics that had periodically swept across continents, crum-
bling ancient social structures and shifting the course of history,
was pinned on microbes—specific, tangible creatures, not foul
odors and evil spirits—lurking in the food, water, soil, and air.

Once the enemy was targeted, it was only a matter of time until
preventive measures, and eventually effective weapons against
specific disease agents, began to be developed. By the turn of the
century, preventive vaccines against diphtheria, typhoid, smallpox, and yellow fever were ending the terror of some of human-
kind's oldest scourges. Antiseptics cut the terrible toll of wound
infections. Sanitation measures and public health crusades

against microorganisms and the insects that spread malaria, typhus, and other infectious diseases greatly lowered the outbreak of epidemics. By the 1930s sulfa drugs were developed, and ten years later, antibiotics, arming physicians with treatments for many bacterial diseases that evaded the preventive measures. The era of therapeutics—specific treatments targeted to specific disease agents—had begun.

The battle has not been won. Infectious diseases, altogether, are still the fourth leading cause of death in the United States, behind heart disease, cancer, and strokes. One reason for the never-ending battle is the genetic flexibility of microorganisms. Flu viruses can mutate and spring virulent new strains on populations vaccinated for a previous flu strain. Bacteria were trading genes among themselves before man knew what genes were. Genes for resistance to antibiotics are quickly passed on from one bacterial type to another, causing the rise of hardy new strains that elude our therapies. New strains now plaguing us include penicillin-resistant gonorrhea and the Gram-negative bacteria that cause widespread infections among hospitalized patients. (Bacteria didn't invent the gene-swapping tactic to elude a man-made threat. We got our antibiotics from some of the natural enemies of bacteria, soil molds.) As soon as one pathogen is controlled, others seem to grab our attention, causing infections such as Legionnaires' disease, acquired immune deficiency syndrome (AIDS), and toxic shock syndrome.

Another reason infectious diseases haven't been completely conquered is that vaccines and modern drugs simply haven't been made available to everyone on earth who needs them. Only 10 percent of the world's children are vaccinated, and five million children die every year from diseases such as measles for which vaccines exist.

Despite its shortcomings, the war on infectious diseases has been the medical miracle of the twentieth century. Life expectancy—which, before Pasteur, had hovered at around forty years—has risen steadily in the United States and Europe to today's average of seventy years.

But with the external causes of disease understood and under increasing control, the spotlight has shifted to another source of suffering and distress: malfunctions in our own internal systems. The thousands of exotic and common ills included in this category—from relatively rare inherited disorders like sickle cell anemia and cystic fibrosis to the major killers, heart disease and cancer—cannot be blamed on outside forces alone. Our vulnerability lies in the molecular machinery of life itself. It is the search for the roots of these diseases that has brought us inevitably to the threshold of gene therapy, treatments designed to alter or supplement the genes an individual was born with.

The targets for treatment lie within each of our cells: the quirks and flaws in the fifty thousand to one hundred thousand genes that shape our development and orchestrate the chemistry of our lives. Current treatments do not strike at the source. We supply insulin to diabetics, transfuse blood into patients with hereditary anemias, and dole out blood-clotting substances to hemophiliacs. We use drugs to strip away copper or toxic wastes that defective cells cannot process and others to damp the pain and inflammation of arthritis. For cancer we try to kill as many malignant cells as possible without killing the patient in the process. By watching our diet, exercise, and work habits, we can cut our risks of falling victim to our inborn vulnerabilities to heart disease, ulcers, or diabetes. Today we can treat the symptoms and control the progress of many genetic and metabolic diseases, but we cannot yet cure the underlying defects. However, just as our battle against infectious diseases progressed rapidly once germs were identified, the possibility of intervening at the genetic level has advanced steadily in the forty years since we found out what genes were.

3. Unraveling Inheritance

And the earth brought forth grass, and herb yielding seed
after his kind, and the tree yielding fruit, whose seed was
in itself, after his kind . . .

—Genesis, 1: 12

Our ancestors understood intuitively that parents have something to do with what their children and grandchildren will look and act like. After all, "like begets like." Cattle give birth to cattle; wheat seeds bear wheat. Ignorant of the principles of evolution as well as of inheritance, early humans still had a notion of the plasticity and changeability of species. They managed to figure out that if one of their sheep or corn plants showed some desirable characteristic, it could often be perpetuated by careful breeding. In a few tens of thousands of years of selective breeding, humans completely remodeled wild species of dogs, cattle, sheep, pigs, horses, and other livestock, and the plants that provide us with grains, beans, other foodstuffs, and fibers. At least since the time of the Greek philosopher Socrates, quoted in Plato's *Republic*, advocates have called for the same sort of "eugenic" selection of human mates to improve our species or "to prevent the human race from degenerating." But exactly what it was that could be selected—beauty, wisdom, character, or good fortune?—and how it got passed along remained a puzzle until our own century.

People seem to have realized from ancient times that certain afflictions run in the family: harelip, dwarfism, feeblemindedness. An individual might be cursed or favored, ugly or fair, robust or crippled, clever or simple, and his ancestry seemed to have a great deal to do with it. When madness struck, when children sickened and died, or when young bodies became twisted into crippling forms, early physicians had no way to tell if the roots of the malady were inherited or not. And they had no cure, no matter what the cause. Until modern science figured out just what is transmitted from parent to child, and how, it was impossible to begin to sort out which elements of the human condition were "in the blood" and which were simply passed along as a consequence of the family's living conditions and social and economic fortunes.

Earlier cultures had no dearth of speculations on how babies are made and why they turn out as they do. The ancient Greeks debated whether the direction of the wind at mating or the temperature of the "seed" determined the sex and appearance of a child. Into the nineteenth century, birth defects were considered punishments or omens, or the consequences of a shock or fright the mother had suffered during pregnancy.

Near the end of the seventeenth century, making use of the newly invented microscope, Dutch researchers discovered that semen isn't just liquid, it carries cells called spermatozoa. One of the microscopists imagined he saw in the sperm the image of a tiny human form he called a "homunculus," preshaped and complete, waiting to be born and to grow. Within the invisible sex cells of the tiny homunculus, he postulated, were tinier homunculi, and so on to infinity—all of future mankind prepackaged like an ever-smaller nest of boxes. Not until 1824 was the female contribution, the egg, discovered. Still, the material by which hereditary traits were carried in egg and sperm remained unknown.

By the mid-nineteenth century, Charles Darwin had formalized the principles of evolution: Random changes, mutations, occur in the material of inheritance, and natural selection acts to preserve those changes that give a creature some advantage in reproducing itself and perpetuating its lineage. Greater hardiness to cold. More efficient processing of some scarce nutrient. Protective coloration to hide it from predators. A pelvic structure that allows it to walk on its hind legs and leaves its front limbs free for carrying. (Human selection in crop plants and livestock, on the other hand, favors perpetuation of inherited characteristics that are useful or pleasing to man, but wouldn't necessarily help the animal or plant to thrive in the wild.) However, even in Darwin's time, nothing was known about the physical units of inheritance that formed the basis for evolution, nor about the laws by which they were shuffled, altered, and passed along.

At about the same time Darwin's work was stirring controversy in England, a monk in Austria was crossbreeding pea plants and keeping meticulous records of the results. In 1866 Gregor Mendel proposed that what creatures inherit is a set of physical particles or "factors" that govern each of their traits. He also outlined the mathematical rules by which these factors are passed from generation to generation. But unlike Darwin's writings, Mendel's work went virtually unnoticed for almost forty years.

It wasn't until the beginning of the twentieth century that Mendel's work was rediscovered and the particles of inheritance he postulated came to be called "genes." Scientists realized that what living things inherit is a blueprint, not a preformed structure. Each new individual is self-assembled.

In the eighty years since, researchers have been mutating, isolating, dissecting, synthesizing, mapping, splicing, and transferring genes in an effort to find out exactly what they are made of and how they work.

By the 1920s it was determined that genes are located on threadlike structures called "chromosomes" found in the nucleus of each cell. When scientists exposed chromosomes in the egg and sperm cells of fruit flies to X-rays, the result was random mutations that changed the physical traits of the next generations of flies—red or white eyes, straight or curly wings, extra body parts, etc. This work proved beyond a doubt that physical units of heredity existed and were subject to change. But what were they made of?

Researchers found out what the physical stuff of heredity was by watching bacteria exchange it among themselves. When harmless bacteria were exposed to certain substances taken from a pneumonia-causing strain, some of them took in these substances and were changed permanently into virulent microbes. They also passed on this virulent trait to future generations. In 1944, scientists identified the material that caused this inherited change—deoxyribonucleic acid or DNA.

Nine years later James Watson and Francis Crick identified the now-famous double-helix, or spiraling zipper, structure of the DNA molecule. Each of the two strands or tapes of the zipper is made up of sugars and phosphate groups, and the interlocking teeth are sequences of four chemical units or "bases": adenine (A), guanine (G), cytosine (C), and thymine (T). The sequence of these bases running along each DNA strand forms a code, a genetic instruction written in a four-letter alphabet. The paired strands in a DNA molecule are complementary—A on one strand is always paired with T on the other, and G with C. It is this complementarity that allows living things to grow and reproduce. The DNA strands are unzipped before the cell divides, and each can thus serve as a template for assembling a new partner.

Chromosomes are long double strands of DNA, alphabetic sequences that contain lengthy stretches of apparent garble interspersed with meaningful sequences we call genes. A gene may be hundreds or thousands of "letters" long, a single genetic sentence. The sentence spells out for a cell a single instruction—how to make one of the thousands of proteins needed to build, support, and operate a plant or animal. It took scientists until 1961 to learn just how the instructions a gene carries are put into operation. Inside the nucleus of a cell, a working copy of the DNA sentence is constructed out of a related material—ribonucleic acid or RNA. This go-between, called messenger RNA or mRNA, carries the instructions into the cytoplasm where they are translated into protein. Triplets of DNA or mRNA letters code for each of the twenty different amino acids that can be strung together to make proteins. CGG codes for arginine, GCC for alanine, and so on. By 1967 the code was completely deciphered.

Mutations can rearrange the genetic sentence, spelling out instructions for a modified protein that may or may not work as well as the original. All the genetic sentences in a cell form a book called the "genome"—the total of all the genetic instructions needed to build and run an organism.

The four-letter genetic code has been found to be the same for all life-forms on earth, from algae to elephants. The number and contents of the genetic sentences in a genomic book determine what sort of creature it will build. Within a single species, such as man, all individuals will share a very similar book, containing fifty thousand to one hundred thousand genetic sentences. The volumes are similar enough that none of us will be born with wheels or need to breathe carbon monoxide instead of oxygen. But not all our genetic sentences are worded the same way. It is the variations in phrasing that are the source of individuality. They will make us men or women, blue- or brown-eyed, tall or short, nearsighted, bald, or prone to allergies or ulcers. They may point us toward scholarly pursuits or soldiering, or ensure that we will never be violin virtuosos or track stars. When the variations are extreme, we call them defects or diseases, and they can leave us retarded or crippled or sentenced to an early death. Our genomes are the range of possibilities within which we and external forces shape our lives.

In the thirty years since the structure of DNA was announced, we have learned to take genes apart letter by letter, deciphering the entire genetic blueprint needed to build a number of small viruses. The three billion letters of the human code will take us longer to read, but we have already deciphered the instructions for some important human proteins, such as insulin and growth hormone. We have learned to build artificial genes that work, even turned the task over to automated DNA synthesizing machines—and we can modify the natural spelling, the sequence of bases, to turn out novel proteins more suited to our needs. We can identify and isolate single genes and map their locations on the chromosomes. And since the advent of recombinant DNA or gene-splicing techniques in 1973, we can chop out and splice together pieces of DNA from any source and insert these hybrid or recombinant DNA molecules into bacteria for mass production by cloning. (Making multiple copies or clones of a molecule is a different process from cloning a whole frog, mouse, or man, as we will see.) As bacteria carrying these hybrid DNA molecules grow and divide, producing a new generation every twenty minutes and a billion descendants in fifteen hours, they generate billions of identical copies of the foreign DNA.

During the mid-1970s, this newfound ability to link up unrelated bits of DNA and give microbes novel genetic abilities caused uneasiness among scientists and often hysterical fears in the general public. Behind the fears about the safety of the work with microbes lay deeper concerns—what came to be called the "Frankenstein factor": Scientists were tampering with the very nature of life, perhaps eventually even human life. The result of the controversy was strict federal regulations on gene splicing. Over the years, however, as the laboratory strains of microbes most commonly used by researchers have proved too domesticated to live in the wild and no "Satan Bugs" carrying novel diseases have emerged, the rules have been eased. By the late 1970s public fears turned to wonder and excitement as the commercial possibilities of recombinant DNA began to dominate the headlines.

The first medical payoffs of genetic engineering have come not from putting new genes into people, but inserting them into bacteria. Bacteria outfitted with foreign genes can be turned into virtual factories, churning out products nature never taught them to make or need. The commercial and medical potentials of the technique became clear in 1979 when bacteria were "taught" to make the hormone that controls human growth. This hormone, previously available only in small amounts and at great cost from human pituitaries collected at autopsy, is injected into children born with hereditary dwarfism to stimulate more normal growth rates. Shortly afterward, human insulin and the human antiviral substance interferon were produced in bacteria. Researchers at universities, pharmaceutical firms, and bustling new genetic engineering enterprises throughout the world are busy outfitting bacteria to produce blood-clotting factors and other scarce human hormones and enzymes as well as human and animal vaccines.

Drugs produced by bacterial "middlemen," however, can't provide us with a way to relieve all our genetic ills. When the problem is lack of a hormone such as insulin that is ordinarily secreted into the blood, injections from an outside source can provide a life-saving treatment. But even with regular injections of insulin, diabetics are vulnerable to heart disease, strokes, and blindness, and they average two thirds the normal life expectancy. Bacteria can't help when our bone marrow turns out defective components and can't assemble healthy red blood cells, nor can they help when toxic wastes build up in our brain cells or when the immune system fails to mobilize any of our complex array of defense mechanisms. To relieve these and hundreds of other hereditary disorders, we must change the activity of our faulty genes—or send healthy foreign genes into the proper cells to do the work our own genes cannot handle.

4. Designer Genes

Perhaps within the lifetime of some of us, the code of life processes tied up in the molecular structure of proteins and nucleic acids will be broken. This may permit the improvement of all living organisms by processes which we might call biological engineering.
　　　　　—EDWARD L. TATUM, 1958 Nobel Prize lecture

As soon as the units of heredity were located, scientists and philosophers began to speculate about how we might escape the genetic endowment bestowed on us by the random pairing of egg and sperm and have a hand in assembling our own blueprints. Dreamers have not limited their visions to correcting obvious genetic defects. Once the code is broken, why not reshape the whole creature?

The late British biologist J.B.S. Haldane—one of the first to propose cloning carbon copies of people—once speculated that someday we might create mutant men bred for special niches: gillbreathers for life in the ocean and "a regressive mutation to the condition of our ancestors in the mid-Pliocene, with prehensile feet, no appreciable heels, and an ape-like pelvis" for life in low gravity on the moon. More recent visions range from a Maine biologist's answer to the population explosion—genetically shrinking all future humans to a height of three feet so they take up less space—to commentator Andy Rooney's personal order for genes to make him tall, thin, and naturally skilled at tennis.

Such speculations touched off a round of soul-searching about the enormous social consequences of "remaking man" that continues today among scientists, ethicists, philosophers, politicians, and government officials. Focusing on the most grandiose or frivolous scenarios tends however to mask just how close we are to the possibility of making specific, planned changes in our genetic inheritance.

A decade ago, transplanting new genes to cure inherited diseases seemed to many scientists almost as remote a goal as designing gill-breathing men. And the things they learned by engineering bacteria were not enough to make therapy in humans feasible. Isolating genes, putting them into a single-celled microbe, and turning them on full blast for maximum production are enormously different from transplanting new genes into a large number of cells in specific human or mouse tissues and getting those genes to work in a controlled way, in response to natural signals. Although the genetic code is simple, human genetics is extremely complex. But the field is moving ahead at dizzying speed.

Invading an animal or a human cell with foreign DNA has proved to be relatively easy. Since 1977 researchers have developed three major techniques for introducing genes into the cells of higher animals, all having been predicted at least twenty years earlier:

—Injecting genes into cells with microscopic needles;

—Letting viruses carry them in;

—Exposing cells to naked DNA in a way that will encourage a few cells to take in the foreign genes and put them to use.

Already a second generation of gene transfer techniques—including the domestication of tumor viruses to serve as "vehicles"—is bringing more precision and efficiency to the process.

The rapid advances have seemed so promising at various points that individual research teams decided it was time to take genetic engineering to humans. The enthusiasm has spawned several widely different attempts to alter the genetic makeup of patients with incurable hereditary diseases:

—In 1970 three young sisters in Germany were deliberately infected with a rabbit virus in hopes the virus would supply their cells with an active gene for an enzyme they lacked. The children's inherited deficiency in a single enzyme had left them doomed to retardation and epilepsy. But the effort by American and European scientists was heavily criticized by their peers and ended with no apparent benefits to the children.

—In 1980 an American physician attempted to transplant healthy genes for hemoglobin production into the bone marrow of two young women in Israel and Italy. The women were suffering life-threatening heart problems and other complications from an inherited blood disorder called beta-thalassemia. This work, too, was cut short quickly by criticism that the physician had jumped the gun scientifically and done the experiments without proper authorization. The women's conditions were apparently not changed by the gene infusions.

—In 1982 a half dozen patients with beta-thalassemia and sickle cell anemia were dosed with an anticancer drug in an effort to activate hemoglobin genes that had been dormant since before birth. The levels of fetal hemoglobin did increase temporarily, but the experimental treatments were not continued long enough to improve the patients' conditions.

The reaction of most laboratory-based molecular biologists to all three attempts ranged from serious misgivings to chagrin. Many physicians working with hopelessly ill patients, however, were glad to see the frontier breached. Gene manipulations in these cases seemed to them no different in spirit or ethics from a heart or kidney transplant. But the decision to move from test tube and mouse experiments to the bedside often involves a tug between desperate medical needs and the desire for a more complete understanding of the possible risks and benefits of a new technique. The political controversy that surrounded the advent of genetic engineering in microbes during the mid-1970s, combined with the uneasiness many people feel about any genetic intervention in humans, has made decisions involving gene therapy even more sensitive than those involving toxic new cancer drugs.

The same rapid progress in the laboratory that has excited clinical researchers and spurred the attempts in human patients has impressed on basic scientists just how much still has to be learned about human genetics. Getting new genes into cells has been relatively easy, but controlling how they work in the alien setting has not been so simple. Recent results in animals, however, indicate we are getting closer. During 1982 two groups succeeded in making permanent, inheritable changes in a physical trait of an animal by an infusion of foreign genes—a feat that had been accomplished before only in microorganisms:

—A line of brown-eyed fruit flies was converted permanently to one of red-eyed fruit flies when the gene for red eyes was injected into fly embryos.

—A strain of giant mice, up to double the normal size, was created by injecting mouse embryos with a hybrid gene—a rat growth-hormone gene spliced to a powerful control signal that normally turns on an unrelated mouse gene.

In both cases the flies and mice passed on their newly acquired genetic traits to new generations of offspring.

* * *

These very visible successes in the genetic engineering of animals have revived the speculations and debate that began with the discovery that DNA carries our physical inheritance: Do we want to tamper with the genetic endowment of human beings at all? If so, should we intervene only to cure dreadful diseases? Or should we try to create "better" people, or people more suited to the lifestyles and habitats of the future? Should the changes we make, for whatever purpose, be limited to the individual's body cells, or should they include the egg and sperm cells so that the changes will be carried into future generations?

Late in 1982 a presidential commission report and testimony before a congressional subcommittee brought pressure for the creation of a new oversight group to monitor developments in human genetic engineering. Manipulations in humans, the commission report noted, will not only help us cure disease but will present "a challenge to some deeply held feelings about the meaning of being human and of family lineage." Both as individuals and as a society, we will have to understand the nature and implications of these new powers if we are to use them responsibly.

II
INHERITED
ILLS

5. The Genetic Lottery

... if we assign only one sort of cause it is not easy to
explain all the phenomena; the distinction of the sexes;
why the daughter is (or can be) like the father and the son
like the mother, and again, the resemblance to remote
ancestors, and further, the reason why the offspring is
sometimes unlike any of these but still a human being; but
sometimes, proceeding along these lines, appears finally to
be not even a human being but only some kind of animal
that is called a monstrosity."

—ARISTOTLE

A mother in Victorian England noticed the frightening symptoms—the deep "staining of the napkins"—only a few days after her child was born. On exposure to air, the infant's urine turned black. The rare condition had been named alkaptonuria, and it fascinated a turn-of-the-century London physician-biochemist, Sir Archibald E. Garrod. Looking into dozens of cases of alkaptonuria, Garrod found that most were easily traced to early infancy, and that the condition usually occurred in several brothers and sisters within the same family. Although the parents were almost always normal, Garrod noted a striking fact: A very large proportion of them were first cousins who had married.

The abnormality was clearly "in the family," but how did it develop? It was popular knowledge that the children of first cousins were apt to suffer an unusual number of ills. However, when Garrod published his study in the British medical journal *The Lancet* in 1902, he rejected the notion that alkaptonuria was a spontaneous defect provoked by the union of two close relatives.

"There is no reason to suppose that mere consanguinity of parents can originate such a condition as alkaptonuria in their offspring, and we must rather seek an explanation in some peculiarity of the parents, which may remain latent for generations, but which has the best chance of asserting itself in the offspring of the union of two members of a family in which it is transmitted," he wrote.

To explain how such a hidden "peculiarity" might surface, Garrod turned to the work of an obscure Austrian monk named Gregor Mendel. Mendel had published the basic laws of inheritance in 1866, but they had gone virtually unnoticed until their rediscovery in 1900. When Garrod applied them to a human abnormality, alkaptonuria, two years later, he outlined the foundations for what was to become the modern science of medical genetics. Like Mendel's principles, however, Garrod's insights were to go virtually unheeded for forty years.

Mendel had gathered his data in pea plants, not people. By picking single, visible traits that seemed to occur regularly in various varieties of peas, he was able to do what no one before had accomplished—to record in painstaking detail the passage of traits from generation to generation. Some varieties of peas were yellow, some green; some were tall, some short. Some had round seeds and others wrinkled seeds. Season after season, Mendel carefully crossbred and planted peas with various traits in the monastery garden, keeping tedious records of the results—counting the wrinkled seeds and the round, the yellow peas and the green. By applying mathematics and logic to his records, the monk laid the groundwork for the science of genetics.

Mendel proposed that each inherited characteristic he had observed was governed by a single physical "factor" or "element"—renamed gene four decades later—that was passed from generation to generation. Several variations of the same factor could exist: The factor for wrinkled seeds must differ from the factor for round seeds. When pea plants carrying different factors are crossed, the hybrid offspring must inherit a factor from each parent. The two governing factors must also be discrete physical units, he reasoned, since they do not mingle and mix, causing any plant to produce both wrinkled and round seeds or something in between.

Mendel had noticed, however, that all the offspring of a cross between wrinkled and round-seeded peas had round seeds. The wrinkled factor (w) seemed to disappear or hide. When the hybrids were crossed with each other, however, the wrinkled trait showed up again in about 25 percent of the second-generation plants. Mendel proposed that the trait for round seeds (R) was somehow "dominant." The first-generation hybrids had inherited both types of factors (R and w), and both were still present, but the R factor prevailed and hid the presence of w. When these hybrids were crossbred with each other, however, their offspring could inherit any of three combinations of factors: a round factor from both parents (RR), a round factor from one and a wrinkled factor from the other (Rw or wR), or two wrinkled factors (ww). Only plants with two wrinkled factors would have wrinkled seeds. Mendel called such traits as wrinkled seeds "recessive," and his observations explained why certain characteristics—or "peculiarities" like alkaptonuria—seem to vanish for generations and later reappear.

Complex traits governed by multiple genes follow more complicated inheritance patterns and show up in a greater range of varieties because of the numerous possible combinations of genes any individual can inherit.

Today we know that the same principles that apply to peas hold true for humans and all other living things. All our physical and biochemical characteristics are governed by one or more genes. Each person has perhaps fifty thousand to one hundred thousand genes—we are not yet sure how many—carried in twenty-three pairs of chromosomes in the nucleus of every cell. Twenty-two of the chromosome pairs have been assigned numbers according to size—chromosome 1s are the largest and chromosome 22s the smallest. In these twenty-two pairs, the chromosomes are alike, so that the cell has a backup copy for every gene. The twenty-third pair are the sex chromosomes. Two matched X chromosomes specify a female. One X paired with a shorter Y makes a male. Half of our chromosomes—one member of each pair—come from our mothers via the egg and half from our fathers via the sperm. (History is full of tales of women cast aside or beheaded for failing to produce male heirs for their husbands. But the mother has only Xs to contribute. It is the X or Y chromosome carried by the sperm that determines the sex of a child.)

Chromosomes are not passed on intact. A child won't get one of the identical chromosome 6s with all the same genes that Great-Grandma had. The genetic deck gets shuffled with every generation through a process called recombination or crossing over. When eggs are being formed in the ovaries, the chromosome pairs in the developing egg cells line up—number 6 from the woman's mother next to number 6 from her father and so on—then the pairs swap lengths of genetic material. Every egg will carry twenty-three chromosomes, each one patched together from lengths of its grandpa and grandma's genes. The same shuffling occurs as sperm is formed in a man. In the genetic lottery that unites a particular egg and sperm, the result will be an individual with a unique assortment of inherited traits.

Garrod realized that Mendel's laws explained for the first time the ancient observation that children of incest or of first-cousin marriages tended to suffer more than the normal share of defects and abnormalities. Despite genetic shuffling, a person receives half his genes from each parent. When that person has a child, he transmits half his genes to the child. The child, therefore, will get roughly one fourth of his genes, in a random assortment, from each grandparent. First cousins who share a set of grandparents are more likely to share some genes in common than are two unrelated people. When the gene happens to be an abnormal recessive like alkaptonuria, every child of that union has a one-in-four chance of inheriting the abnormal gene from both parents and thus showing the trait.

"Such an explanation removes the question altogether out of the range of prejudice, for . . . it is not the mating of first cousins in general but of those who come of particular stocks that tends to induce the development of alkaptonuria in the offspring," Garrod noted. "For example, if a man inherit the tendency on his father's side his union with one of his maternal first cousins will be no more liable to result in alkaptonuric offspring than his marriage with one who is in no way related to him by blood."

This application of Mendel's laws to a human abnormality was important in itself, but Garrod took his theories several steps farther. First, he suggested that alkaptonuria might not be the only biochemical peculiarity man was heir to. Albinism, the lack of pigments that color the eyes, skin, and hair, and cystinuria, a condition that causes urinary tract crystals and stones, seemed to follow a similar pattern of incidence.

"All three conditions above referred to are extremely rare and all tend to advertise their presence in conspicuous manners," Garrod observed. "May it not well be that there are other such chemical abnormalities which are attended by no obvious peculiarities and which could only be revealed by chemical analysis?" People differ not only in "various tints of hair, skin, and eyes," but in their reactions to drugs and their "degrees of natural immunity against infections," he wrote. Garrod envisioned for the first time the chemical uniqueness of each individual:

"If it be, indeed, the case that in alkaptonuria and the other conditions mentioned we are dealing with individualities of metabolism and not with the results of morbid [disease] processes the thought naturally presents itself that these are merely extreme examples of variations of chemical behavior which are probably everywhere present in minor degrees and that just as no two individuals of a species are absolutely identical in bodily structure neither are their chemical processes carried out on exactly the same lines."

Garrod next explored how our chemical individuality comes about. Other researchers had already identified the substance that made an alkaptonuric's urine turn black. It was alkapton, or homogentisic acid. But why was it there? If the condition was not the result of some newly acquired disease process, then it had to be an inborn "sport" or novelty, Garrod wrote. Alkaptonurics excreted this substance, he proposed, because their bodies lacked an enzyme that is normally used to metabolize it or break it down. (It took until 1958 for researchers to confirm this predicted enzyme defect.) The abnormality seemed harmless. "However, regarded as an alternative course of metabolism, the alkaptonuric must be looked upon as somewhat inferior to the ordinary plan, inasmuch as the excretion of homogentisic acid in place of the ordinary end products involves a certain slight waste of potential energy," he concluded. (The blackening of the urine is harmless, but we know today that this chemical abnormality also affects the joints, leading to severe arthritis in midlife, and carries a risk of premature heart attack.)

Garrod then made a more prophetic deduction: The reason patients lacked the enzyme to break down alkapton was that they were missing the normal form of a gene. (Of course, he didn't call it a gene, since the hereditary units weren't named as such until a few years later.) It was the first time anyone had stated what has come to be known as the "one gene, one enzyme" concept: The function of a gene is to carry the instructions for making a single enzyme or other protein the body requires. Genes, therefore, control the biochemical processes of the body. People are different chemically and metabolically because they are different genetically. Alkaptonuria and other extreme variations that Garrod eventually labeled "inborn errors of metabolism" were the result of changes or mutations in specific genes.

The hypotheses had little impact in Garrod's own lifetime. It remained for American scientists George Beadle and Edward

Tatum to prove the one gene, one enzyme concept in the 1940s with their experiments in pink bread mold. By exposing the lowly mold to radiation and mutating its genes, the two researchers were able to produce errors of metabolism at will, nailing down the link between a creature's genes and its biochemistry. In 1958 when the two researchers won a Nobel Prize for their work, Tatum noted in his Nobel lecture that the results of their experiments implied that ". . . each and every biochemical reaction in a cell of any organism, from a bacterium to man, is theoretically alterable by genetic mutation . . ."

It is these inherited mutations, these inborn errors, that we classify today as genetic diseases.

For a half century after Garrod published his findings, the science of genetics progressed rapidly, but it had little impact on the practice of medicine or on the plight of patients with genetic disease. Genes were so named and found to be carried in chromosomes (although it wasn't until 1956 that we learned the correct number of human chromosomes). Genes of fruit flies and other humble creatures were mutated and mapped by order, location, and function. Genes were discovered to be made of DNA, and the storied double-helix structure of DNA itself was unveiled, ringing in the modern era of molecular biology.

During the same period, however, only a few more diseases were added to Garrod's list of inborn errors of metabolism. American physician James V. Neel observed during the years after World War I that sickle cell anemia, a blood disorder that affects blacks almost exclusively, was inherited in the same recessive pattern as alkaptonuria. Norwegian physician Ivar A. Folling found in 1934 that a certain type of mental retardation was the result of a defective enzyme in the liver. The condition, called phenylketonuria or PKU, is also caused by a recessive gene.

After World War II, the pace of medical genetics began accelerating rapidly. Today more than 3,500 diseases, each the result of a specific variation in a single gene, have been cataloged. Physicians have come to realize there is no "normal" genetic makeup. People are hormonally and metabolically different, and not all the differences are harmless. Each of us carries an accumulated burden of peculiarities and flaws, including as many as five to ten potentially harmful recessive genes. Fortunately, most of us inherit only one copy of each of these "bad" genes, and we have a healthy backup copy from our other parent to protect us from its harmful effects.

More recently we have begun to sort out subtler variations in our genes that, as Garrod noted, affect our natural immunity against infection and also contribute to the development of cancer, heart disease, arthritis, diabetes, and other ailments that are individually much more common than those officially cataloged as genetic defects.

The time lag between basic science and clinical medicine is also closing—too quickly, some researchers feel. Decades passed before the earlier work with peas, fruit flies, and bread mold gained any relevance for physicians treating retarded, crippled, or dying children. But medical genetics has followed molecular biology into the laboratory. When today's genetic engineers use gene splicing and gene transplant techniques to double the size of mice or change the eye color of fruit flies, medical researchers are immediately aware of the possibilities. They are already using many of the basic scientist's techniques to pinpoint the genes responsible for various inherited disorders and to probe one step beyond what Garrod had prophesied: to identify the specific molecules within each defective gene that keep it from functioning normally. It is this precise knowledge of molecular anatomy that physicians will need to begin treating genetic disease at the level of causes instead of symptoms.

6. Inborn Errors

Nature is nowhere accustomed more openly to display her
secret mysteries than in cases where she shows traces of
her workings apart from the beaten path; nor is there any
better way to advance the proper practice of medicine than
to give our minds to the discovery of the usual law of
Nature by careful investigation of cases of rarer forms of
disease. For it has been found, in almost all things, that
what they contain of useful or applicable nature is hardly
perceived unless we are deprived of them, or they become
deranged in some way.

—WILLIAM HARVEY, 1657

Something clicked in Linus Pauling's mind one winter evening in
1945 during a dinner in New York. Pauling, who would win the
1954 Nobel Prize in chemistry, was listening to a physician describe his work with a disease called sickle cell anemia. In the
arteries the patient's hemoglobin-filled red blood cells, richly
loaded with oxygen, are normal and disk-shaped. When these
cells have delivered their oxygen to some outlying tissue, however, and are traveling back to the heart through the veins, they
distort into long crescent-moon or sickle shapes.

". . . the idea occurred to me that sickle cell anemia was a
molecular disease, involving an abnormality of the hemoglobin
molecule determined by a mutated gene," Pauling wrote in 1970
as he recalled the occasion.

Back at the California Institute of Technology, Pauling and his
team set out to demonstrate the suspected flaw in the sickle hemoglobin molecule. The researchers subjected various samples of
hemoglobin to an electric field, a technique called "electrophor-
esis," to see how each behaved. As Pauling had theorized, there
was a difference. The sickle protein moved slower when electric
current was applied than normal hemoglobin did.

Medical researchers using Mendel's classical statistical methods had already observed that the disease was inherited as a re-

cessive trait. Parents of sickle cell anemia victims—carriers who had one gene for the sickle cell trait—apparently produced enough normal hemoglobin to oxygenate their tissues and did not suffer from the disease. A person had to inherit two sickle genes for the chronic anemia to show up. Pauling's studies confirmed the inheritance pattern for the disease. Hemoglobin from carriers proved to be about half normal and half sickled, each type moving at a different pace in the electric field, betraying the presence of one normal and one sickle gene.

For the first time, researchers were able to "see" at the molecular level proof of what Garrod had suggested in the first decade of the century and Beadle and Tatum had demonstrated in the 1940s: A mutant gene produces a change in the molecular construction of a protein. Human genetics became an experimental science instead of a discipline that relied solely on statistical data to track and probe inherited disease.

Pauling's landmark study was published in 1949. Seven years later, German American biochemist Vernon M. Ingram dissected the hemoglobin molecule, using electric current to spread the chemical building blocks of the protein molecule—the amino acids—out across sheets of filter paper. Comparing what he called the "fingerprints" of the two molecules, normal and sickled, Ingram found a tiny difference: Out of the 574 amino acids in a molecule of hemoglobin, one was different in the hemoglobin of sickle cell patients. Where a molecule of glutamic acid was supposed to be placed in the chain, valine sat instead. The substitution made the sickle hemoglobin slightly less acid, and this was the reason it had moved slower than the normal protein when Pauling had applied an electric current to it.

One swapped amino acid out of 574 hardly seems serious, but the effects are devastating. An estimated one American black in every ten and perhaps 20 percent of African blacks carry the recessive gene for sickle cell. One in every five hundred black children inherits two genes for sickle cell and is born with sickle cell anemia. The disease also occurs, but less commonly, among people of Mediterranean and Indian descent. About fifty thousand individuals in the United States suffer from the disease. Their rigid and distorted red blood cells wedge in small blood vessels, causing clots and blockages. When the clogging occurs in the spleen, it provokes a painful sickle crisis. The spleen cannot perform its job of making antibodies to defend the body against invaders, and the patient is left vulnerable to potentially lethal infections. Half of all patients die before the age of thirty. Today researchers understand the defect down to the level of atomic interactions, but despite this precise knowledge, we still have no cure. Inherited blood disorders like sickle cell top the list of suggested candidates for gene therapy.

The catalog of genetic diseases contains more than three hundred other observed abnormalities of the hemoglobin molecule in addition to sickle cell anemia, including a group of disorders called thalassemias found primarily among Mediterranean peoples. Hemoglobin is just one of the tens of thousands of protein products needed to build, support, protect, and operate a human body, each one coded for by a gene. Theoretically there may be at least one inherited disease corresponding to every hormone, enzyme, and other protein produced by our bodies.

At Johns Hopkins University in Baltimore, geneticist Victor McKusick has become the archivist of genetic defects, compiling genetic diseases identified by researchers around the world, storing them in a computer databank, and publishing every five years an updated volume entitled *Mendelian Inheritance in Man*. As of February 15, 1983, his computer contained 3,436 mutation sites or disease loci, 1,660 confirmed, and the rest still tentative. The number of identified genetic diseases is actually larger than this, however, because a single genetic locus may be subject to many possible defects. The 300 known hemoglobin abnormalities, for instance, are included as only two entries among the 3,436 (one entry each for the alpha- and beta-globin genes, the genes for the two types of proteins needed to assemble hemoglobin molecules in adults).

In addition to these single gene disorders, about a hundred gross defects in chromosomes have been identified in human diseases—extra or missing chromosomes or chunks of certain chromosomes that have broken away from their rightful sites and settled on the ends of other chromosomes.

In disorders that affect red blood cells, the culprit protein is fairly obvious. Red cells contain little but hemoglobin and water. Of the other inherited diseases identified so far, however, researchers have pinpointed the defective gene or its deficient protein product only in about 275 cases.

The thousands of known genetic disorders include a huge array of consequences. Some defects are so severe that 95 percent of the occurrences end in spontaneous fetal abortion. Others have impacts as mild as nearsightedness or the inability to use certain drugs. The consequences of other disorders range from mental retardation, cancer, heart defects, and clotting problems to color blindness. The impacts may come during fetal life, childhood, or middle age.

Each of the cataloged genetic diseases is relatively rare by itself, but together the disorders are a leading cause of sickness, suffering, disability, and death in infancy and childhood. Their proportion has increased as improvements in obstetrical care, nutrition, sanitation, and control of infectious diseases have dramatically reduced other causes of infant disease and death.

—Four to 6 percent of infants are born with clearly defined genetic diseases, chromosome abnormalities, or severe congenital defects.

—When disorders like Huntington's chorea that reveal themselves only later in life are included, an estimated 10 percent of the population carries a potentially handicapping inherited abnormality, perhaps fifteen to twenty million individuals in the United States.

—Inherited diseases and genetically influenced birth defects account for 20 percent of all infant deaths in the first year of life in the United States, second only to the number caused by birth injuries and premature birth. They are the second leading cause of death in the one-to-four age group and the third in the fifteen-to-nineteen age group.

—An estimated 40 to 60 percent of all human fetuses conceived are lost to miscarriage or spontaneous abortion. At least half of those spontaneously aborted in the first three months of pregnancy have chromosome abnormalities.

—Inherited defects account for 30 percent of pediatric admissions at major hospitals and 13 percent of adult admissions. The costs for medical and institutional care range in the hundreds of millions of dollars. A 1972 study estimated that hereditary disease and congenital defects caused the hospitalization of 1.2 million people in the United States at a cost of more than $800 million.

Genetic disorders can be grouped according to the effects on the patient, the type of molecular defect, or the pattern by which the disorders are inherited.

One effect is hearing loss. Victor McKusick's catalog lists more than 125 genetic sources of deafness, including inherited diseases known primarily for their effects on the kidneys, nervous system, or eyes. One of the most devastating impacts of faulty genes is mental retardation. An estimated two thirds of the cases of severe retardation are attributable to inherited defects, the rest to a wide variety of nongenetic causes such as infections, birth injury, brain damage, and drug effects. Approximately five hundred inherited disorders, ranging from defects in single genes to the presence of extra chromosomes, are known to lead to some degree of mental crippling.

Of all the known causes of mental retardation, the most common is Down's syndrome, first described in 1866. This defect is sometimes called mongolism, a rather prejudicial and unfitting label, because those afflicted have slanted eyes and other slightly Oriental features. Affected children can also have a number of other symptoms: creases that run straight across their palms instead of branching, a wide separation between their great and second toes, fissured tongues, and heart defects. Their I.Q.'s average about half of normal, although the range of retardation runs all the way from mild to profound. In 1959 French geneticist Jerome Lejeune was studying the chromosomes of Down's children and discovered a telltale abnormality. They have forty-seven chromosomes instead of forty-six—an extra copy of chromosome 21 and all the genes it carries. Which extra genes are responsible for the various symptoms, and how, are questions geneticists are still working on.

A less common genetic disorder called Prader-Willi syndrome results in a very different assortment of symptoms: mild retardation, voracious appetite, and extreme obesity. Researchers studying the cells of these patients have found pieces of genetic material swapped between chromosome 15 and various other chromosomes, suggesting that something important may have gotten lost in this abnormal chromosome shuffle.

A condition among males called fragile X is considered second only to Down's syndrome among the identifiable causes of mental retardation. A break in the tip of the patient's X chromosome causes one end to dangle, giving the defect its name.

Defects in single enzymes can also lead to retardation. Children born with phenylketonuria, or PKU, for example, are deficient in a single gene product, a liver enzyme called phenylalanine hydroxylase needed to convert an amino acid called phenylalanine. If phenylalanine intake is not sharply reduced by a special diet during childhood while the brain is developing, severe retardation results. Individuals with a disorder called Lesch-Nyhan syndrome also inherit a defect in a single enzyme with a polysyllabic name—hypoxanthine guanine phosphoribosyl transferase, or HGPRT. The result is mental retardation combined with cerebral palsy and a strong compulsion for self-mutilation.

Tay-Sachs disease victims lack an enzyme known as hexosaminidase A. Without it, certain lipids or fats build up in the brain and other nerve cells, leading to retardation, paralysis, and death at an early age.

Since any broad impact such as retardation can result from a multitude of causes, scientists often find it more useful to group hereditary diseases according to the nature of the molecular defect involved. As we have just seen, however, the molecular defect has been uncovered in only about 8 percent of the known single-gene diseases.

Nearly 225 of the enzyme deficiencies that Garrod labeled inborn errors of metabolism have been pinpointed, including PKU, Tay-Sachs, and Lesch-Nyhan syndrome. Missing or defective enzymes, often with obscure and unpronounceable names, usually produce their effects by tripping up one of the myriad biochemical cycles in the body, causing a foul-up down the line and crippling some vital life process. This can damp the immune system (adenosine deaminase deficiency), leave the joints loose and the skin abnormally stretchable (Ehlers-Danlos, or "rubber man" syndrome), bring on fits of madness (porphyria), or cripple the body's DNA-repair system and lead to skin cancer (xeroderma pigmentosum, or XP).

About fifty hereditary disorders tracked to the molecular level affect nonenzymatic proteins such as hemoglobin (sickle cell anemia and the thalassemias), hormones (growth hormone defects), clotting factors (hemophilias A and B), and immunoglobulins (agammaglobulinemia).

Hereditary defects can occur in "receptor" molecules on the surfaces of cells that are supposed to grab other molecules from the blood and admit them into the cells. Defects in genes for various types of receptors can cause a range of impacts from heart attacks to feminization of male testicles. When receptors for cholesterol-carrying molecules called low density lipoprotein (LDL) are reduced or absent, the result is a potentially deadly buildup of blood cholesterol. The disorder is called familial hypercholesterolemia, and it is the most prevalent genetic disorder yet identified, affecting one in every five hundred Americans.

Other known molecular defects can lead to malformed sex organs, foul up the transport of metals and nutrients, or cause certain specialized types of cells in the bones or immune system to fail to develop.

For families with a history of certain inherited disorders, or for those who have already borne a child with a genetic defect, the identification of the inheritance pattern of a disease is important. Often this is the only information they have to use in family planning. In the vast majority of genetic defects, from cystic fibrosis to Huntington's chorea, no tests are available yet to determine whether a fetus has inherited the disease. A known inheritance pattern provides at-risk families with at least a gambler's choice— the odds that genetic roulette will produce another affected child.

Some of the most severe genetic disorders are passed along in the manner described earlier as recessive traits. Many recessive abnormalities are concentrated in ethnic or regional groups: for example, sickle cell anemia in blacks; cystic fibrosis carried by one in every twenty-five white Americans; Tay-Sachs among Eastern European Jews and an isolated group in Nova Scotia; the thalassemias in Mediterranean, Middle Eastern and Far Eastern peoples, and blacks; PKU and homocystinuria among the Irish; porphyria among white South Africans; and PKU and a type of dwarfism among the Amish of Pennsylvania. If two carriers of the same recessive gene marry—something more likely to happen in closely knit groups or isolated areas—each child they bear will have a 25 percent chance of getting two "bad" genes and inheriting the disease, a 50 percent chance of inheriting only one abnormal gene and becoming a carrier like his parents, and a 25 percent chance of receiving two normal genes. Recessive disorders usually begin their impacts early in life, often before birth.

The nature of sexual inheritance leaves males subject to a special category of recessive abnormalities that range from color blindness to Duchenne muscular dystrophy, Lesch-Nyhan syndrome, and the bleeding disorders hemophilias A and B (sometimes called the "disease of kings" because England's Queen Victoria, through her children, passed along the trait to royal families in Germany, Spain, and Russia). These are the X-linked or sex-linked disorders, carried by genes on the X chromosome. Since women have two X chromosomes, they can carry these recessive traits on one chromosome but not be affected by them. Men have only one X chromosome, however, and so are unprotected against the effect of the recessive gene. A mother who is a carrier has a fifty-fifty chance of passing along the gene to every son.

Ironically, progress in medical genetics may bring some new risks for women. They can be affected by X-linked diseases if they inherit the trait from both a carrier mother and a father with the disease. In the past, men affected by serious X-linked disorders seldom survived long enough or were fit enough to reproduce. That may change as better treatments for genetic diseases become available.

Diseases caused by dominant genes are usually less severe than those caused by recessive ones and often do not appear until midlife. Dominant disorders usually have variable penetrance, meaning that not everyone who inherits the gene manifests the disease, and those who do experience a wide range of severity. Only one dominant gene is needed to produce an effect. If every carrier of an abnormal dominant gene were retarded, crippled, or killed as quickly as those who inherit two severe recessive genes, the dominant gene would have died out with its victims long ago. Every child of a parent with a dominant genetic disease has a fifty-fifty chance of inheriting it. A certain proportion of cases result from new mutations and arise in families with no history of the disorder.

Hereditary brachydactyly—very short fingers—is a dominant trait. So is neurofibromatosis—"Elephant Man's disease"—a fairly common disorder that afflicts eighty thousand Americans. Neurofibromatosis may cause dozens of lumpy, unsightly tumors under the skin and sometimes internal tumors, but prompt surgery can prevent serious complications. Familial hypercholesterolemia is a dominant disorder predisposing its victims to premature heart attacks.

A striking exception among dominant disorders in its severity and penetrance is Huntington's chorea, the degenerative disease of the nervous system that felled folk singer Woody Guthrie. A person carrying the abnormal gene usually lapses in middle age into a progressive physical and mental deterioration that can go on for a decade or more before ending in death. The disease appears after many unsuspecting victims have already had children. There is currently no test that can tell whether a child of a Huntington's victim has also inherited the disease. Since the onset of symptoms is sometimes delayed until the age of sixty or beyond, this uncertainty can last a lifetime. As many as one hundred thousand people in the United States may be affected.

Chromosome abnormalities usually arise from new mutations and are not passed along in a Mendelian inheritance pattern. The risk of recurrence in relatives is low. Most chromosomal disorders that are recognized in early life or childhood are associated with mental retardation and obvious malformations in physical structure or organ systems.

Some chromosome defects, particularly those involving abnormal numbers of sex chromosomes, have less visible impacts and may go unnoticed. Men with XYY, an extra Y chromosome, are taller than average and often have lower-than-average intelligence (an I.Q. in the 70s or 80s). What other traits they inherit is controversial.

In the mid-1960s, chromosome studies on men in mental institutions and prisons showed there was a greater proportion of XYY males in such places than in the general population. Many of these men had a history of violence. Controversy raged for a decade over whether the trait was predictive of aggression and criminal tendencies. In 1976 a Danish study of men in the general population gave some firmer answers: Forty-two percent of the XYY males located had criminal records, compared with 9 percent of males examined and found to have normal chromosomes. But except for a case of wife beating, none of the XYY males had committed violent or aggressive crimes. Most, in fact, had perpetrated such dull-witted mischief as burglaries carried out while people were at home. The researchers concluded that XYY males do have some tendency toward antisocial behavior, although not necessarily aggressive acts. Remember, though, that more than half the XYY males identified were apparently leading upstanding lives.

There is another group of genetically influenced disorders that is not as easy to track through a family tree as dominant and recessive traits. The diseases and birth defects in this group, including cleft lip and palate and many types of congenital heart disease, do not follow the rules. They seem to involve one or more genes interacting with each other and sometimes with one or more environmental factors as well. These defects are dubbed "multifactorial." The odds of an individual in an afflicted family inheriting such a disease are harder to figure, but they are much lower than the odds for classic recessive and dominant diseases. The risk to relatives of afflicted individuals is in the 1 to 4 percent range, although this percentage rises if more than one family member is affected. Studies in twins have made it clear, however, that genes play a key role in determining who will be affected.

Identical twins are clones, split from a single egg, shaped by the same genes. From ancient times people have regarded twins with a certain awe, marveling that two individuals could be so strikingly alike. Twins' physical similarities go beyond outward form and feature, often involving traits as personal as blood pressure and pulse, headache and sleep patterns, dream cycles, vision, and weight problems. But the similarities don't tell the whole story. Twins often have striking differences, too, despite being genetic duplicates. It is these differences that are helping researchers to probe the limits of genetic influence and to unravel the interplay of genes with environmental factors, in the womb or in the outside world.

If one identical twin is born with a club foot, for example, there is a 30 percent chance the other twin's foot will be similarly twisted. This is a ten times higher risk than that faced by nonidentical twins, who are no closer genetically than any set of brothers and sisters. Although clearly genes must play a role in determining who is at risk, genes alone do not cause club foot. Remember that 70 percent of the time, the second twin escapes the fate of his or her genetic duplicate.

Cleft lip and palate bring similar odds: If one twin has a defect, the identical twin will share it 40 percent of the time. Again, that is ten times the risk faced by a child whose nontwin brother or sister has a cleft lip or palate, and four hundred times the risk faced by the population at large. Still, 60 percent of the time, the second twin escapes the defect.

This unpredictability doesn't usually show up in genetic diseases passed on in families according to the dominant and recessive inheritance patterns described by Mendel. In most of those cases, if the abnormal genes are there, so is the disease to some degree. Even the most favorable prenatal environment will not cause the abnormal genes of a hemophiliac to start ordering up normal blood-clotting factors.

Researchers estimate 20 percent of birth defects can be blamed on abnormal dominant or recessive genes acting alone. Ten percent are caused by abnormalities in chromosomes. The other 70 percent apparently occur only when a certain combination of genetic and environmental factors interact. Besides clefts and club feet, this category includes spina bifida, in which the spinal cord protrudes from the lower back, and anencephaly, in which most of the brain and skull are missing. Some of these multifactorial disorders can be pinned largely to environmental causes—drugs like thalidomide, alcohol, radiation, and viruses such as the one that causes rubella, or German measles. But even the effect of these teratogens (agents that cause fetal malformations) seems to be modulated by genetic susceptibilities.

In recent decades researchers have come to realize that genes have a much wider influence on our health than causing these individually uncommon inborn disorders and birth defects. The interplay of genes and environmental factors—from viruses and cigarette smoke to radiation, asbestos, and diet—is responsible for some of our most common afflictions: diabetes, heart disease, cancer, mental illness, and many of the chronic and debilitating diseases of later life.

No one has yet isolated the gene, or set of genes, involved in most of these diseases, or pinpointed all the environmental factors that trigger or modulate them. But, as is the case for cleft lip, the evidence for genetic involvement is clear in the statistics and the frequency with which certain disorders tend to crop up in families.

If one twin develops insulin-dependent diabetes, for instance, the other has a fifty-fifty chance of getting it, too. In the case of non-insulin-dependent diabetes, the risk goes up to more than 90 percent that if one identical twin develops the disease, both twins will suffer it. The odds are 45 percent that both twins will develop schizophrenia if one twin has it. Certain types of cancer and heart disease, which account for almost two thirds of the deaths in the United States, have long been known to cluster in families.

The interaction of genes and environment in disease should come as no surprise. The same subtle mix of factors shapes most of the normal characteristics of living things. A grove of cedars growing near the sea, constantly buffeted by ocean breezes, may grow shrublike and crooked, branched only on the leeward side. Inland, out of the wind, genetically identical cedars will grow into straight and substantial trees, evenly branched all around. A child may inherit the tendency to grow to a taller-than-average height and all the genes needed to carry out that tendency. Yet poor nutrition, infectious diseases, or certain drugs can thwart this inborn programming.

In the past, even when afflictions seemed to run in the family, it was often impossible to tell whether it was because of physical heredity or the life-style and fortunes of the clan. Environmental factors acting alone can often produce results that look just like those produced by abnormal genes: Retardation can be caused by injuries or oxygen deprivation at birth instead of faulty genes or chromosomes; anemia by iron deficiency rather than defective hemoglobin; depression or psychosis by grief, stress, or drugs rather than inborn chemistry.

Statistics and studies of twins and adoptees have given scientists their first clues, confirming a role for genes in many of our most common diseases. Now the techniques of molecular biology are allowing them to begin sorting out the interwoven roles of genes and environmental factors and to devise interventions at the proper level, whether this means changing family life-styles or replacing faulty proteins—or, eventually, the genes that produced them.

7. Inborn Vulnerabilities

Only a few percent of the population suffer from a serious
disease of physical or mental development of the kind
usually associated with faulty genes, but many common
diseases have an important genetic component, some
individuals being inherently more susceptible than others.
If we prorate the genetic factor in disorders like diabetes,
atherosclerosis and schizophrenia, we estimate that at least
25 percent and perhaps half, of our total health care burden
has a genetic basis.

—JOSHUA LEDERBERG, 1970

To maintain its own integrity, a body must recognize itself. The
soldier cells and defensive molecules that patrol the blood and
tissues need signals to help them tell friend from foe. Without
such a system our defenders might attack our brain, liver, or muscle cells while letting bacteria, viruses, and other foreign invaders
roam at will. To prevent this confusion, our genes provide each of
us with a unique self tag, a molecular monogram worn on all our
cells. This monogram is known as our HLA type. From what scientists are learning about the system, the HLA types may soon
provide us with another service: advance warning of inherited
vulnerabilities to a wide range of diseases, from diabetes and hay
fever to multiple sclerosis.

HLAs—human leukocyte antigens—are white blood cell
types. (A leukocyte is a white blood cell.) The four human red
blood cell types—A, B, AB, and O—were discovered around the
turn of the century. That discovery made safe blood transfusions
possible for the first time, but physicians found that some patients
who were given foreign blood of a type that matched their own
still suffered the chills and fever of transfusion reactions. Researchers had no explanation until the late 1950s, when white
blood cell types began to be identified. The discovery of the HLA
system, however, has had a much broader impact than increasing
the safety of blood transfusions. It has paved the way for the typ-

ing and cross matching of tissues that have made kidney, heart, and other organ transplants so widely successful. The more HLA markers an organ donor and a recipient have in common, the less likely the recipient's body will recognize the transplant as foreign and reject it.

The HLA markers are products of a cluster of genes on chromosome 6 known as the major histocompatibility complex, or MHC (*histo* means tissue). Despite its name, the MHC has turned out to be more than a signal service that determines tissue compatibility. It is the master control region for the immune system, our internal security agency that recognizes and defends us against viruses, bacteria, and other disease-causing agents. The functions of the MHC gene cluster are so complex and far-reaching the region has been dubbed a "supergene."

Researchers in the mid-1960s had injected mice with leukemia virus and found that animals carrying certain HLA markers (they are called H2 types in mice) were twice as likely to come down with the disease as were mice with other HLA types. The work sparked others to begin looking for statistical relationships between human HLA types and diseases. By 1970 a link between HLA and a form of cancer, Hodgkin's lymphoma, had been confirmed. Ten years later the list of apparently associated diseases had climbed to more than sixty. The diseases included hypertension, coronary artery disease, recurrent herpes virus infections, tooth decay, multiple sclerosis, hemochromatosis (in which the body stores too much iron), myasthenia gravis (which causes nerve and muscle degeneration), chronic active hepatitis, insulin-dependent diabetes, psoriasis, systemic lupus erythematosus (a chronic inflammatory disease of the skin, joints, and other systems), asthma, hay fever, schizophrenia, manic depression, and a half dozen types each of cancer and arthritis.

That first decade of work on HLA and disease obviously produced tantalizing clues about the genetic factors behind some of our major health problems. But the clues, the statistical links, weren't strong enough to be useful to physicians or individuals trying to predict their personal risks. One of the strongest links, for instance, is the relationship between the HLA marker B27 and a chronic spinal arthritis called ankylosing spondylitis. About 90 percent of those who suffer from the disease carry the B27 marker. Obviously a small percentage of people who get ankylosing spondylitis don't have the marker. And the vast majority of people who do have it—about 9 percent of the population—will never get the disease. Which B27 carriers are at risk? And what environmental factors are needed to trigger the susceptibility into a full-blown disease?

These questions couldn't be answered in the 1970s, partly because the HLA markers and the system used to define them were too crude. Modern genetic engineering techniques are quickly refining the process, identifying new categories of markers and subdividing older types. The result should be a much greater understanding of the organization and function of the MHC and a much more precise linkage between HLA types and various diseases.

Typing has traditionally been done by reacting a person's blood sample with 100 to 150 different antibodies that recognize and react with specific HLA markers on the cells. (Antibodies are proteins produced by the immune system in response to a substance, an antigen, it recognizes as foreign.) Labs currently get their antibodies from the blood of women who have borne several children. The women's blood carries low levels of antibodies against certain HLA markers on their babies' cells—the markers inherited from the father and thus recognized as foreign by the mother.

This painstaking process of extracting and purifying antibodies from human blood and cross-reacting them with blood samples is the method scientists have used since the 1950s to separate and define the HLA types. The first sets of HLA markers identified were the A, B, and C factors, now known as Class I antigens, each available in eight to thirty varieties. But these turned out to be just the tip of the iceberg. By 1980 the Class II HLA markers, D and DR, had been identified. Many diseases associated earlier with the ABC antigens were found to have even stronger links to D-DR types. In the past few years, the number of known marker sets has proliferated further. Next to DR, researchers found another site called DC. Then a new series of antigens called SB was identified. Each type of antigen is apparently the product of a different gene or cluster of genes in the MHC. A person inherits one variety of each of the antigens from each parent—A9, B27, C2, DR3, etc.—a unique constellation of markers for each individual.

By 1980 researchers had gone beyond the HLA types defined by blood studies and had begun to look directly at the genes that produce the markers. In the past few years, as scientists have isolated and cloned the genes of the MHC, they have found that the DR and SB sites contain multiple genes for each of the two types of protein chains (alpha and beta) needed to assemble each antigen. This indicates that each person may carry a number of DR markers rather than the two (one from each parent) that the blood tests have been defining.

The antibodies used in the past have not been sensitive enough to distinguish between all these subtypes of gene-produced HLA markers. Today researchers are developing ways to "read" the DNA directly, or type cells using highly precise "monoclonal" antibodies manufactured by genetic engineering techniques (more about these methods later). The result is that all the associations between HLA and disease are being restudied in hopes of finding tighter linkages with precise new markers. Carl Grumet at Stanford University in California, for example, has been working with monoclonal antibodies that subdivide HLA B27 types into different variants, investigating whether one of these subtypes is more associated with ankylosing spondylitis than the other. Hugh McDevitt's team at Stanford and several other labs around the country are working on subdividing the DR3 category, which is strongly associated with diabetes and several other diseases. About 85 percent of people who develop insulin-dependent diabetes have the DR3 or DR4 markers, compared with 30 percent of the general population.

Some of the disease associations identified in the past may not survive this second round of winnowing. Other new diseases will be added to the list. The big question for the 1980s is how the HLA markers are related to disease resistance or susceptibility. Some diseases may turn out to be associated simply because the genes involved are located near the MHC on chromosome 6 and tend to get passed on from generation to generation in their company. But scientists suspect other diseases will fall into a tighter, causal relationship.

One group in particular is the so-called autoimmune diseases. These include insulin-dependent diabetes, multiple sclerosis, rheumatoid arthritis, myasthenia gravis, systemic lupus erythematosus, and the adrenal gland disorder Addison's disease. Increasing evidence indicates these diseases are results of mistaken identity, or double cross, in which the body's defense system turns against its own tissues. What might trigger such a double cross? Several theories have been put forth: A virus or other foreign antigen may mimic one of a person's self tags, confusing the body's security force and causing it to attack both the virus and its own tissues. Or viruses or other invaders may alter a cell's HLA monogram in such a way that the immune system no longer recognizes it as self. A third possibility is that microbes or trauma might also unmask a marker on certain tissues that normally remains hidden and thus isn't recognized by the body's defenders as self.

HLA linkages are confirming the involvement of genetic factors in these autoimmune diseases and others, but much work is still needed to pin down which genes are at work and how. An even tougher task will be to identify the environmental triggering agents that actually bring on the diseases in susceptible people.

The very nature of the HLA system has led researchers to believe that the MHC genes play a direct role in these disease susceptibilities and in our ability to fight off major infections. The A, B, and C markers are involved in guiding the body's killer T cells (a type of white blood cell) to virus-infected cells that need to be destroyed. The DR and SB markers, found only on certain immune cells, play a role in triggering the production of defensive antibodies by B cells, another type of white blood cell.

Why do the HLA markers come in such wide variety? One theory is that variations evolved because each made the body more resistant to a different type of infectious disease. Evidence indicates some HLA markers do a better job than others of calling attention to certain invaders and coordinating the body's defenses against them. And some markers tend to occur more frequently among specific ethnic groups and in certain geographic areas, suggesting they may offer more protection against major infectious diseases found in those regions.

Research on the HLA system promises to shed light on the nature of the host factors that make some individuals and groups more vulnerable than others to certain infectious pathogens. Why do some people with genital herpes virus infections suffer frequent and painful recurrences while others never have a second attack? Why doesn't everyone exposed to the tubercle bacillus suffer from tuberculosis? Why do some people end up sniffling and congested every winter and others seldom suffer colds?

Finally, some researchers believe the so-called supergene may be involved not only in determining what each of us is likely to suffer and die from but also how soon it will occur. Life-span in various species correlates directly with the efficiency of several critical systems that repair DNA damage and protect the cell machinery from the toxic by-products of everyday chemical reactions. There is some evidence that the MHC genes regulate these repair-and-protection processes. Roy L. Walford of the University of California at Los Angeles believes that the immune system, and thus the genes of the MHC, will turn out to be the "pacemaker" that regulates the human aging process.

Diabetes

Genes, viruses, and mistaken identity are all implicated in the group of diseases called diabetes. Being overweight and out of shape don't help your chances either. Ten million Americans—5 percent of us—have diabetes. Only six million have been formally diagnosed as diabetics, however, and another 600,000 are added each year. An estimated four million people in the United States are undiagnosed diabetics. Another five million are considered on the borderline.

The common factor in all diabetes is the inability to handle glucose (sugar) properly. The body—specifically the liver, muscles, and fatty tissues—needs this sugar to burn for energy or to store. In normal individuals the hormone insulin, secreted by islet cells in the pancreas, helps the tissues pick up and use sugar from the blood. But diabetics either don't produce insulin or their bodies don't make proper use of what they do produce. In either case, sugar collects in the blood while the cells go hungry. The body then begins to cannibalize itself, burning fats and proteins instead of sugar. But the cells cannot burn fats completely without some of the by-products from the burning of sugar. Toxic acid substances from the incomplete burning of fats build up in the body, and unless insulin is supplied, the patient lapses into a diabetic coma and dies.

Diabetes is divided into two categories: Insulin-dependent—also called juvenile—diabetes affects about 10 percent of diabetics. The other 90 percent have non-insulin-dependent, or adult-onset, diabetes. Twin studies have shown that both types have a genetic basis, but scientists are still trying to pinpoint the nature of the inherited vulnerabilities.

Insulin-dependent diabetes is the most severe form of the disease. It appears suddenly, usually around the age of twelve, although it can begin in the first year of life. Patients produce little or no insulin themselves and usually need one or more injections of insulin a day to survive.

Strong evidence indicates that insulin-dependent diabetes is an autoimmune disorder. In 80 to 85 percent of newly diagnosed cases of the disease, the patients' blood has been found to contain defensive antibodies targeted against their own insulin-producing islet cells. This could be caused by an inherited quirk in the diabetics' immune systems. Or it could be that the insulin-producing cells have been so damaged by other insults—chemicals or viruses—that the body doesn't recognize them anymore.

Even though insulin-dependent diabetes seemingly appears suddenly, some studies show it may take years to develop. George S. Eisenbarth and his colleagues at the Joslin Diabetes Center in Boston reported in 1983 that they had followed a pair of identical twins and a set of triplets for seventeen years after one child in each set developed diabetes. Eventually, the second twin and one of the two remaining triplets began to turn out antibodies against their insulin-producing cells. Insulin manufacture declined gradually until both individuals finally developed the disease. During those seventeen years, the third triplet developed neither antibodies nor diabetes. Clearly some other factor besides genes is involved, since the triplets are genetically identical.

That other triggering factor may be a virus. The second triplet had already begun to produce antibodies when the study started. Over the next eight years, his insulin production gradually declined and his blood-sugar levels rose, but not to the point where he became diabetic. Then he came down with infectious mononucleosis, a disease caused by a strain of herpes virus. Suddenly his insulin production dropped dramatically. Within five months he was an insulin-dependent diabetic.

Earlier studies with animals genetically susceptible to diabetes had shown that infection with common human viruses can destroy insulin-producing cells and trigger the disease. In 1978 virologist Abner L. Notkins of the National Institutes of Health in Bethesda, Maryland, found the first direct evidence that viruses can bring on diabetes in humans. His team isolated a virus called Coxsackie B4 from the pancreas of a ten-year-old boy who had died during a severe onset of insulin-dependent diabetes. When the virus was injected into genetically susceptible mice, it destroyed insulin-producing cells and brought on diabetes. Several similar cases in children have been reported since then.

Even though researchers don't know yet which gene (or genes) is involved in this type of diabetes, or how, dozens of labs around the world are working to pin down a more precise linkage between the disease and a specific HLA marker. A precise marker would allow physicians to use a simple blood test at birth to determine which children had inherited the susceptibility. These children could be closely watched for the development of antibodies against their own insulin-producing cells. In some animal studies, immunosuppressive therapy—the same kind used to damp the immune response and prevent the rejection of transplanted organs—has succeeded in preventing the development of diabetes after antibodies to insulin-producing cells appeared. Results with this type of preventive therapy in humans have been mixed so far.

There is even stronger evidence for genetic predisposition in the second major type of the disease, non-insulin-dependent diabetes, although no markers have yet been found that predict reliably who is most vulnerable. Remember that if one identical twin gets this type of diabetes, the other is almost certain to develop it, too, and people with this type of diabetes often have a family history of the disease. But again, genes aren't the whole story. Obesity seems to be the major triggering factor in non-insulin-dependent diabetes.

Ninety percent of diabetics have the non-insulin-dependent type, and they usually develop it after the age of forty. Few of these people suffer from a lack of insulin. Many, in fact, have normal or higher-than-normal blood levels of insulin. The trouble is that they are insensitive and unresponsive to the insulin they produce. This is called insulin resistance, and there are several possible reasons for it.

An insulin molecule does its work by binding to a special receptor on the surface of cells and triggering some signal or series of signals inside the cell that cause the cell to transport sugar from the blood through its outer wall and process it. Insulin resistance in an individual may be caused by too few insulin receptor sites on the cells, some defect in the receptors that causes them to lose their affinity for insulin, or any of several possible abnormalities in the way the cell itself responds to the insulin or the signals it triggers.

Whatever the nature of the insulin resistance, it cannot be blamed solely on faulty genes. Studies in animals and humans have shown that obesity can reduce the number of insulin receptors and induce insulin resistance in the tissues. Eighty to 90 percent of non-insulin-dependent diabetics are overweight. Obesity can bring on diabetes in genetically susceptible individuals, and further weight gain can aggravate the disease. Exercise, weight loss, and sulfonylurea drugs can increase the body's sensitivity to insulin and control blood sugar in many patients. But when patients fail to exercise regularly or fail to lose weight, physicians often have to resort to high doses of insulin to try to overcome their insulin resistance and bring blood-sugar levels down.

An intense search is on for genetic markers that can predict who is at risk for developing this type of diabetes. It has been known for several years that an extra fragment of DNA is present at the front end of the human insulin gene, and that its length varies in different individuals. Various groups of researchers have tried to associate the fragment lengths with non-insulin-dependent diabetes, but the results have been conflicting. It remains to be seen whether the extra piece of genetic material means anything, and if so, whether it is part of the cause of the disease or just a marker that indicates a genetic susceptibility.

Heart Disease

The impact of defective genes on the heart can be profound. At least fifty specific genetic disorders are known to cause congenital malformations, defects in the structural materials that form blood vessels and heart valves, diseases of the heart muscle and pericardium (the sac that surrounds the heart), abnormalities in rhythm and heartbeat, heart tumors, high blood pressure, and coronary artery disease. The vast majority of heart diseases, however, are believed to be caused by multiple genetic and environmental factors.

Heart diseases fall into four major categories. First, coronary heart or artery disease—atherosclerosis or "hardening of the arteries" that supply blood to the heart muscle itself—affects 4.5 million Americans and causes 40 percent of deaths in the United States. Hypertension or high blood pressure affects thirty-seven million Americans, one of every four adults. About 25,000 babies are born with congenital heart defects in this country each year. And rheumatic heart disease affects nearly two million people.

Coronary heart disease, which leads to heart attack, has long been known to run in families, and genes are a contributing factor. If one female identical twin suffers a heart attack, there is a 44 percent chance the other will, too. For female fraternal twins, the risk is 14 percent. Researchers suspect there are probably numerous genetic and environmental factors that can interact to cause coronary heart disease. Among the most important seem to be factors that increase levels of fats in the blood, a primary contributor to atherosclerosis.

At least 20 percent of heart attacks occur among individuals who suffer specific single-gene disorders that prevent their bodies from handling fats or lipids. These diseases are called hyperlipidemias and include familial hypercholesterolemia, familial hypertriglyceridemia, multiple lipoprotein-type hyperlipidemia (also called familial combined hyperlipidemia), and type 3 hyperlipidemia. The first three disorders are dominant, the last is recessive.

Lipoproteins are particles that transport lipids such as triglycerides and cholesterol through the blood. Each lipoprotein is a complex of a water-soluble protein and a fat molecule. There are four main types of lipoproteins: LDL (low density lipoprotein) is the main carrier of cholesterol. Chylomicrons and VLDL (very low density lipoprotein) carry most of the triglycerides in blood. And HDL (high density lipoprotein) carries the "good" cholesterol fraction, apparently removing it from tissues. Hyperlipidemias result from genetic defects that interfere with the production, degradation, or structure of one of these plasma lipoproteins and thus affect the fat content of the blood. Each of these disorders predisposes its victims to premature atherosclerosis.

(A study reported in 1981 by the University of Cincinnati's Charles J. Glueck suggests that these disorders may also account for most unexplained strokes in children: "We speculate that familial lipoprotein abnormalities may mediate cerebrovascular arteriosclerosis and thus predispose children to ischemic strokes.")

At least seven other recessive genetic disorders can also leave their victims vulnerable to premature coronary artery disease. Researchers speculate, but have no hard evidence yet, that carriers of some of these recessive genes—which could include one in every thirty to one hundred Americans—might also be predisposed to heart attacks.

Since environmental and nutritional factors influence the course of coronary artery disease in genetically vulnerable individuals, medical geneticists are trying to identify the "susceptibility genes" and sort out how they interact with such factors as cigarette smoking, lack of exercise, and excess dietary fats. The information will eventually lead to development of better preventive measures and drugs.

About 5 percent of all congenital heart malformations are caused by chromosome abnormalities, and another 5 percent by single mutant genes. The other 90 percent of birth defects involving the heart result from an interplay of genes and environmental factors.

When the cause is a specific inherited disorder, the heart defect is often only one among a whole cluster of symptoms. The majority of chromosome abnormalities, for instance, include heart defects among their array of possible impacts. About half of Down's syndrome patients have congenital heart disease along with mental retardation and the other characteristic symptoms. Among women with the most common form of Turner's syndrome (only one X chromosome), 35 to 50 percent have heart defects. In contrast, defects caused by the interaction of multiple genes and environmental agents usually occur separately.

Environmental factors like rubella, fetal alcohol syndrome, and teratogenic drugs are the primary causes of congenital heart defects in a small percentage of cases. However, not all babies who are subjected to these risk factors during fetal development are damaged. Genetic predisposition apparently plays a role even in these cases.

Multiple genes and environmental factors all seemingly interact to set the blood pressure level—the force of the blood against the vessel walls—in each individual and maintain it at a stable point. The brain coordinates and oversees operations in several organ systems that work together to maintain this steady head-to-toe blood flow and pressure whether an individual is sleeping, sitting at a desk, or running in a marathon. The heart can affect blood pressure by the speed and force of its beating, the blood vessels by constricting or narrowing, and the kidneys by regulating the amount of salt and water in cells and thus the force required to move blood through stiff and swollen tissues. If a defect or upset occurs at any point in the system, the brain must signal the other stations to compensate and restore the balance or the result will be hypertension.

Some single-gene disorders—porphyria, neurofibromatosis, and Fabry's disease—include hypertension in their constellation of symptoms. But most scientists believe hypertension can result from a wide variety of causes.

Untreated hypertension adds to the workload of the heart and arteries and can lead to stroke, heart attack, or kidney failure. Several classes of drugs that dilate the blood vessels, flush water from the tissues, or alter the release or receipt of chemical messengers in the central nervous system can help lower blood pressure in specific individuals. Restriction of salt intake and weight loss can also help some people. Geneticists are searching for markers that may indicate which individuals are vulnerable under what specific circumstances, information that will eventually allow people to plan personal programs for preventing high blood pressure.

Mental Illness and Alcoholism

For centuries people have observed that madness tends to run in families, but how it gets passed along is still the subject of hot debate. Is demented behavior learned from sick parents, or do abnormal genes carry the seeds of insanity? Throughout the twentieth century, various researchers have applied statistics to family lineages to try to track schizophrenia and pin down Mendelian inheritance patterns that would show the affliction is rooted in the genes. The results have been muddled and subjective, partly because it is hard to classify whether a person is insane or not and to what degree. Proponents on both sides of the "nature versus nurture" debate seem to agree that there is a great deal of oddity, if not overt madness, in the families of schizophrenics. These borderline cases have come to be lumped under subjective classifications such as schizoid personalities. But the nature versus nurture question remains: Does a child inherit madness from his parents or learn it in the home?

Identical-twin studies in the 1950s and 1960s won points for the proponents of genetic influence. If one twin is diagnosed as a schizophrenic, there is a fifty-fifty chance the other will be. If a schizoid personality is included in the diagnosis, the chances rise to 80 to 90 percent that the second twin will be affected. In nonidentical twins, who share the same environment but only about half their genes in common, the risk of sharing chronic schizophrenia is about 10 percent—the same as that for any non-twin brother or sister of a schizophrenic. That risk rises to around 50 percent if schizoid conditions are included.

The existence of a genetic factor is clear, but its influence obviously isn't absolute. If it were, identical twins would always share the same fate. Some environmental factors must be involved in determining who eventually gets sick and who doesn't. The nurture proponents have continued to champion the psychological atmosphere within the family as the key, contending that a good upbringing can override whatever genetic predispositions exist. Examining these influences separately within an afflicted family is impossible, however, since parents may provide both "bad" genes and a "sick" environment.

In the 1960s researchers realized that Western social practices had already set up a laboratory in which the impacts of genes and family environment could be studied independently—adoptive homes. The landmark study in this field was published in 1971 by an American and Danish group: Seymour Kety, Paul Wender, David Rosenthal, Fini Schulsinger, and Joseph Welner. The group traced the fate of Danish adoptees born to either schizophrenic or nonschizophrenic parents and raised in adoptive homes. Other research groups have conducted similar studies, and the general results have been the same: Upbringing did *not* seem to play a key role.

More children born of schizophrenic parents turned out to be schizophrenic, too, than were children born of normal parents, no matter what kind of environment they were raised in. In various studies, 10 to 20 percent of the children of schizophrenics developed the illness themselves—up to 45 percent if both parents were schizophrenic—and another 30 to 40 percent displayed schizoid personalities. In contrast, few or none of the children of normal parents in the studies were afflicted.

When researchers looked at children of schizophrenics who were raised by their own sick parents, they found a startling result: These children seemed to have no worse risk of being diagnosed schizophrenic after being raised in a "sick" home than if they had been raised in normal homes. And in the uncommon instances where a child of normal parents was adopted by people who turned out to be schizophrenics, the child had no greater chance of developing into a schizophrenic in the "sick" home than did children of similar heritage adopted into normal homes.

Overall, the genetic predisposition seemed to take its course largely unaffected by the psychological environment; and a "sick" upbringing seemed to produce no illness in individuals who were not genetically vulnerable.

Adoption studies strengthen the evidence for genetic predisposition to chronic schizophrenia. But they leave unanswered the question of what evironmental factors *are* involved in the disease if psychological influences are not. Without insane parents or the domineering mothers and passive fathers of popular myth to blame, what outside forces *do* help to bring on madness? Various studies have implicated factors ranging from viruses to brain injuries in the eventual development of schizophrenia. Studies around the world have shown that children born in late winter and spring have a greater risk of developing the disease (one hypothesis links this to seasonal increases in viral infection). Complications at birth such as prolonged labor also seem to increase the risk of madness.

The causes may be easier to pin down when the nature of the disease is better understood. Schizophrenics—three million in the United States alone—typically display flat personalities, inappropriate emotions, and disordered thinking, and some may suffer delusions and hallucinations, such as hearing disembodied voices. In recent years scientists have turned their efforts to uncovering a suspected biochemical abnormality, some fundamental defect in the workings of the schizophrenic's brain that could be inherited or caused by brain damage, viral infection, or other factors.

Psychoactive drugs introduced since the late 1950s effectively damp the delusions and unrealistic thinking in eight out of ten schizophrenics, and this has convinced most researchers that schizophrenic behavior itself stems from chemical defects in the brain. Theories are plentiful. One is that the symptoms of the disease are caused by an overabundance of a brain neurotransmitter called dopamine, a chemical that carries signals across the gaps between nerve cells. Drugs used to treat schizophrenia block the actions of dopamine.

Other theories suggest there may be a surplus of natural morphinelike substances in the brain, a deficiency of one of the body's hormonelike chemicals called prostaglandins, or an abnormality in any of several other brain neurotransmitters besides dopamine. Using radioactive tracers and sophisticated new scanning instruments, researchers are probing the patterns of activity inside schizophrenic brains in search of possible structural abnormalities.

Oddities in schizophrenics are easy to find. But examining people who are already ill for evidence of a biochemical or structural defect can be misleading. There is no way to tell whether the aberration is a cause or a result of the disease.

It is possible that several abnormalities may work together to cause schizophrenic symptoms, or that different causes may lead to the same result. Some researchers believe schizophrenia is actually several diseases with multiple causes, just like diabetes, heart disease, and cancer. The goal of much current research is to find reliable biochemical or genetic markers that will permit early diagnosis and individualized treatment of patients.

Since the early 1970s, strong evidence for a genetic connection has also been found in the other major forms of mental illness—depression and manic depression—as well as for hysteria, childhood hyperactivity, antisocial or psychopathic personality, and alcoholism. A whole range of nongenetic factors, from a love-starved childhood or social pressure to birth trauma or brain injury, may also cause or contribute to some of these disorders, although these factors are harder to pin down.

Even suicide, which is often an end point of schizophrenia or depression, may sometimes have a biological basis. Adoption studies have found higher rates of suicide among the blood relatives of adopted schizophrenics and depressed patients than among the adoptive families that raised them. In 1981 researchers at the National Institute of Mental Health and the Karolinska Institute in Stockholm, Sweden, reported they had found an apparent "suicide factor" that could be measured to determine which depressed patients or individuals who have tried unsuccessfully to commit suicide are likely to try again.

The factor is a chemical called 5-HIAA, which can be detected in spinal fluid and provides an indirect measurement of certain chemical messengers in the brain. Marie Asberg of the Karolinska Institute noted that low 5-HIAA levels found in suicidal patients do not seem to be the result of their mental illness. Instead they may be associated "with an increased vulnerability, possibly genetically determined, to a range of psychiatric disturbances and to suicidal behavior," she reported.

All of us at times react with grief or unhappiness to tragedies and disappointments in our lives. This type of reactive depression usually has an obvious, nongenetic cause. But millions of people suffer from depressions that seem to have no relation to the ups and downs in their lives and fortunes. These disturbances have various names—organic, biological, vital, or endogenous depression. When periods of depression alternate with irrational highs or mania, the disorder is called manic depression. Psychiatrists group all these conditions under the category of mood disorders, which strike an estimated seven million Americans each year. Adoption and family studies patterned after those carried out for schizophrenia have found a genetic basis for mood disorders, too.

If an identical twin suffers from manic depression, the chances are 50 to 80 percent that the other twin will develop it. The risk among nonidentical twins or among brothers and sisters of a patient with manic depression is about 10 percent. In general, people with a family history of mood disorders have a significantly greater risk of becoming depressed than the rest of the population, even if they are raised in adoptive homes free of mental illness. The notion of a biological cause for these disorders is also strengthened by the fact that drugs introduced in the past two decades bring improvement to 70 percent of depressed patients.

As in schizophrenia, geneticists are busy trying to find imbalances in brain chemicals or to locate genes that might be involved in mood disorders. One of several biochemical markers developed in the late 1970s to diagnose and categorize biological depression is the dexamethasone suppression test, or DST. The body's response to dexamethasone is an indirect measure of certain chemical activities in the brain. Although the test definitely is not foolproof, some clinicians think it can be valuable in helping to establish a diagnosis in confusing cases, especially among children and the elderly in whom biological depression is often mistaken for "a stage she or he is going through" or senility. Other scientists believe the test is not reliable enough for general clinical use and should be used only for research.

Such biochemical tests share the same drawback as biochemical abnormalities identified in schizophrenia: It is impossible to tell whether the imbalances are a result of the disease or part of the cause.

In 1981 Lowell R. Weitkamp of the University of Rochester, New York, and Harvey Stancer of the University of Toronto, Canada, reported finding an HLA marker that seems to signal susceptibility to depression. But as in previous studies linking HLA markers with a predisposition to schizophrenia or manic depression, the finding has been controversial.

In adoption studies of antisocial or psychopathic personalities, researchers have also found firm evidence for the influence of heredity. Both the biological and the adoptive families of diagnosed psychopaths were studied along with those of adoptees who did not turn out to be psychopaths. Only the blood relatives of the psychopaths had a higher-than-normal rate of antisocial personalities. Other evidence, however, suggests that birth trauma, brain damage, and perhaps even an impersonal institutional upbringing can result in the development of antisocial personalities.

Studies in the United States and several other countries have shown consistently that a person's chances of becoming an alcoholic increase with the number of alcoholic relatives he or she has, the severity of their alcohol problems, and the degree of genetic closeness between that person and the alcoholic relatives. Children of alcoholics have higher risks even if they are raised from birth in nonalcoholic adoptive homes. Children with no alcoholic relatives, however, show no increased vulnerability even if they are raised by alcoholics.

Marc A. Schuckit of the San Diego Veterans Administration Hospital and the University of California at San Diego is trying to find out just what it is that alcoholics pass on to their children that predisposes them to alcoholism. In order to avoid confusing the effects of the disease with the actual causes, he is looking for biochemical oddities in high-risk individuals who are not yet alcoholics. Schuckit has found that young men with a family history of the disease appear to respond to alcohol in a chemically different way than those without such a background. His studies have shown that the blood of these individuals contains significantly higher levels of a toxic alcohol breakdown product, acetaldehyde, than the blood of the men without a family history of alcoholism after both groups have drunk the same amount of booze. In other experiments Schuckit has found that the offspring of alcoholics seem to require more alcohol than other people before they feel the same high.

In another approach, he uses an electronic apparatus to measure how alcohol affects one particular brain wave. Early results indicate that the changes in the brain wave in response to alcohol might vary with a person's family history of alcoholism.

Alcoholism, like cancer, diabetes, and possibly even schizophrenia, may turn out to be a collection of diseases with several possible causes. The hope—as with other disorders that involve inherited vulnerabilities—is to develop tests that will provide advance warning to individuals with a high risk of becoming alcoholics. Heavy drinking is almost a rite of passage from adolescence to adulthood in our society, and family, social, and religious influences all play strong roles in determining our lifelong drinking habits. The influence of genetic factors may become most apparent, Schuckit believes, in the mid-twenties to thirties age group when the average person begins to back off from his youthful excess but the "prealcoholic" continues to drink heavily no matter what impact the drinking begins to have on his career or personal life.

8. Genes and Cancer

You may feel disappointed that I have not mentioned any
single lines of progress in medicine which will flow from
the broadening of the foundation of biology. I give you a
free hand. You may wish for anything: a cure-all for
cancer, a mastery of mutation, an understanding of
hormone action, or a cure for any of the diseases you have
especially in mind. None of your wishes need remain
unfulfilled, once we have penetrated deep enough into the
foundations of life. This is the real promise of medicine.
—ALBERT SZENT-GYORGYI, 1963

We barrage our DNA with daily insults: charbroiled steaks, too
much sun, and deep breaths of cigarette smoke and smog. Some
of the insults strike home, shotgunning our genetic instructions
with random breaks and lesions, and keeping an army of repair
enzymes busy patching and splicing the double helix. Most of the
time our defense-and-maintenance staffs handle the assaults well.
The minor nicks and mishaps get fixed. The major hits weaken
and kill a smattering of our hundred trillion cells, but we endure
these submicroscopic ordeals with no discomfort.

Occasionally, however, an insult eludes the system and sets in
motion the process we call cancer. In this case a cell doesn't die,
and its damage doesn't get repaired. Instead the mutation somehow confers new powers on the cell, triggering exuberant and uncontrolled growth. The cell turns outlaw, dividing wildly and
elbowing out its neighbors, to the eventual sorrow and often the
destruction of the body in which it resides.

For decades cancer researchers have tried to figure out what is
different about the mutations that lead to cancer. In 1981 they
found the key: We are born with "hot spots" in our DNA, vulnerable genetic switches that apparently can be triggered to cause
cancer when carcinogens or viruses happen to strike them. A few
of us may even be born with the switches already tripped. Scientists dubbed these hot spots "cancer genes," or "oncogenes"

from the Greek *onkos*, meaning mass or tumor. Their discovery may be the first step toward understanding the molecular events involved in human cancer and designing new strategies for early detection and treatment of the disease responsible for one of every five deaths in the United States.

Oncogenes did not evolve just to plague us. When they are found in normal cells instead of cancerous ones, they go by the unwieldy and somewhat misleading name of "proto-oncogenes" (*proto* means original or primitive), misleading because these genes don't lie dormant like time bombs waiting for the random chance to run wild. They must play some very basic role in the normal growth or operation of cells, and thus in life itself, although researchers have just begun to uncover what that normal role is. The ubiquitous presence of proto-oncogenes is a testimony to their importance. They have been found in all normal vertebrate cells and in some invertebrates, from rats and birds to fruit flies, indicating that nature has found it advantageous to hang on to them throughout eight hundred million years of evolution. By 1983 nearly twenty proto-oncogenes had been found. Most researchers expect the final total will fall between thirty and fifty, mixed in among the fifty thousand to one hundred thousand genes in each of our cells.

Cancer genes exploded into the headlines in 1981, and since that time work on them has been advancing at dizzying speed, thanks to the tools of recombinant DNA and our ability to transfer genes from one type of cell into another. However, the story began much earlier, in the long, frustrating search for viruses capable of causing human cancers. Back in 1910 Peyton Rous of the Rockefeller Institute, New York, announced he had found a virus that caused cancer in chickens, launching the modern field of tumor virology. Plenty of animal tumor viruses were found in the following decades, but none in man. By the 1960s, convinced that viruses must play an important role in human cancer too, researchers began an intensive search for human tumor viruses. When none were found, interest in the theory eventually waned. But some scientists continued to work with animal tumor viruses in the hope they could learn something about how normal cells become transformed into malignant ones.

In the early 1970s, with genetic engineering techniques becoming available, virologists began trying to isolate the genes that gave tumor viruses their power to transform animal cells. An oncogene called *src*, isolated from Rous's chicken virus, was one of the first identified. By the late 1970s, Michael Bishop and Harold Varmus of the University of California at San Francisco, George Vande Woude and Edward Scolnick of the National Cancer Institute (NCI), and others had found about a dozen viral oncogenes. From the beginning, virologists had suspected these oncogenes were not really viral genes at all. Viruses are known to pirate genes occasionally from the cells they infect, and perhaps these oncogenes had been kidnapped long ago from normal animal cells. Then, driven by strong viral signals, perhaps the genes turned on transforming powers they would not have exerted in their normal home.

As early as 1969, Robert J. Heubner and George J. Todaro of NCI had proposed the theory that all normal cells contain within them the seeds of cancer—genes that could cause cancer when triggered by various agents. In the mid-1970s virologists, looking for the origin of their viral oncogenes, set out to see if this theory was true. In 1978 a research team led by Bishop and Varmus demonstrated that all vertebrate species, including humans, carried proto-oncogenes strikingly similar to the cancer genes of viruses. Using gene-splicing techniques, Vande Woude and Scolnick hooked up a viral control region to one of these proto-oncogenes from a normal cell and found that the hybrid gene could turn normal cells malignant.

It was a neat demonstration of how viruses had acquired their cancer-causing abilities, but it brought little public attention. Viruses, after all, had never been shown to cause human cancer. Other scientists were already at work on another line of research that seemed more directly related to humans—the search for cancer-causing genes in human tumor cells. (All this work moved slowly from the mid- to late 1970s because of tight federal restrictions on cloning tumor viruses or cancer genes in bacteria. The rules were relaxed beginning in 1979, and the work took off rapidly.)

To find out if there were genes in human tumor cells that had the power to cause cancer, researchers had to have a way to remove the DNA from tumor cells, insert it into normal cells, and watch the result. A number of labs set to work using a gene transfer technique called "DNA transformation" (see Chapter 16), which had been developed in 1977. The answers came quickly. Researchers took naked DNA from human cancer cells, inserted it into noncancerous cells growing in a glass dish, and found it would turn the other cells malignant. When these malignant cells were injected into a mouse, they formed tumors. In the spring of 1981, Robert A. Weinberg of the Massachusetts Institute of Technology (MIT) and Geoffrey M. Cooper of Harvard University announced the verdict: Oncogenes are present in human tumor cells.

It took less than a year to search through the total DNA of the cancerous cells and isolate the responsible fragment of DNA without which no malignant transformation could have taken place. Three groups—headed by Weinberg at MIT, Michael Wigler at Cold Spring Harbor Laboratory in Long Island, N. Y., and Mariano Barbacid at NCI—announced the isolation of the first human oncogene from bladder cancer cells. Within a short time, genes from lung, colon, and breast cancer cells and from leukemias and lymphomas had also been isolated. By early 1983 the tally of human oncogenes had reached fifteen.

Techniques used thus far have detected oncogenes in about 15 percent of the tumors tested, but that percentage includes cases of almost every type of human cancer, from gall bladder, liver, and pancreas carcinomas to a rare embryonal tumor. And each oncogene does not seem to be limited to causing a single type of cancer. For example, the same oncogene has been found in unrelated lung and embryonal cancers, and three different oncogenes have been isolated from various colon cancers.

The grand convergence of cancer gene research came soon after the isolation of the first human oncogene in 1982, bringing together the results of animal virus and human tumor work: Human oncogenes were found to be virtually identical to viral oncogenes. This means human oncogenes from tumors are almost identical to the proto-oncogenes from normal human cells. Almost overnight, the seeds of cancer became one of the hottest areas of research.

Not everyone shared the excitement. Tumor virologist A. Harry Rubin of the University of California at Berkeley contended the oncogene theory had "the feeling of conforming to the latest fashion, currently that of molecular biology."

However, David Baltimore of MIT notes that it is unlikely to be a coincidence that these oncogenes are present in both human tumors and tumor-causing viruses.

"There's little doubt in the minds of researchers that oncogenes have some fundamental role in human cancer," Baltimore says. "We believe they are fundamental genetic units in the under-standing of what differentiates tumor cells from normal cells."

If so, the next question is: How does a normal, law-abiding proto-oncogene get converted into an oncogene? And once turned outlaw, how does it set in motion the process that leads to cancer? By producing too much of its product at the wrong time and place? Or perhaps by producing an abnormal product? Baltimore points out that there seem to be at least five possible ways a proto-oncogene can go bad.

1. The first way is to be kidnapped by a virus. A cancer gene carried by a virus seems to be locked into the outlaw mode, either because the virus keeps it turned up and producing at abnormal levels or because the virus dropped, swapped, or otherwise modi-fied a piece of the gene it stole. Viruses carrying oncogenes can infect and cause cancer in a wide range of animals, but so far no one has found a case of this type in humans.

2. A virus that does not carry its own oncogene can apparently trigger cancer by invading a normal cell and accidentally turning the cell's own proto-oncogene bad. The virus might sit down by chance next to the gene and interfere with its activity levels, or it might break into the gene itself, mutating it. When the avian leu-kosis virus infects the cells of a chicken and inserts itself into the DNA near a proto-oncogene called *c-myc* (the *c* indicates the nor-mal cellular form of the gene; *myc* designates the oncogene form), the result is a type of lymphoma. Since the virus enters the DNA at random sites, cancer is a rare event, and most chickens con-tract only an infection.

A similar event may set in motion some human cancers. After all the decades of searching, scientists have firmly linked only one class of human cancers to a viral cause—T-cell leukemias and lymphomas. These diseases are rare in the United States but com-mon in Japan and parts of the Caribbean. Robert Gallo of the National Institutes of Health first isolated the virus—human T-cell leukemia-lymphoma virus, or HTLV—in 1980, and with the help of Mary E. Harper of the Agouron Institute in San Diego, is mapping where the viruses locate in the cells of patients who de-velop the cancers.

In 1982 Gallo isolated a second apparent human tumor virus, this one from hairy cell leukemia, and his lab is continuing to search for viruses in patients with diseases like Hodgkin's lymphoma and childhood leukemia. These blood and lymphatic system cancers seem to occur in clusters of people, indicating that they could be spread by frequent or prolonged contact with a virus which, like HTLV, is not highly contagious.

Certain other viral infections are associated with increased risk of cancer, although no cause-and-effect relationship has ever been proved. The herpes viruses that cause cold sores and venereal disease, for instance, are associated with a higher risk of cervical cancer. These viruses can turn rat cells cancerous in the lab, and herpes "footprints" can often, but not always, be found in the tumor cells of cervical cancer patients. The same kind of evidence of a viral visit can often be found for cytomegalovirus in a rare cancer called Kaposi's sarcoma. Kaposi's sarcoma was relatively unknown until it began to strike with unusual frequency in recent years among homosexuals and other victims of acquired immune deficiency syndrome, or AIDS. In another case Epstein-Barr virus is present in 98 percent of patients with African Burkitt's lymphoma, a virulent facial cancer of young children, and yet it has never been established as a cause of the disease.

Researchers are studying the possibility that these viruses may play some sort of hit-and-run role in the cancers they are associated with. Denise A. Galloway of the Fred Hutchinson Cancer Research Center in Seattle notes that the viruses could accidentally trigger the activity of proto-oncogenes or even cause mutations in them. All the herpes viruses that infect humans are known to have the ability to cause chromosome damage. The chain of events leading to cancer might then continue even if the virus left the affected cell.

3. Proto-oncogenes may go out of control in a cell when their numbers become too large, a phenomenon called "amplification." A normal cell may contain one or two copies of the gene, for instance, and a cancer cell fifty copies. Two research groups looked at diseased blood cells from a patient with promyelocytic leukemia and found sixteen to thirty-two copies of the *myc* oncogene.

4. The first human oncogene to be isolated, one called Harvey *ras* taken from human bladder cancer cells growing in culture, acquired its malignant powers by the simplest mutation possible: a change in a single one of the six thousand chemical units or bases that form the normal gene. This is the kind of mutation that chemicals known to cause bladder cancer—such as the

arylamines used in the plastics, dye, and rubber industries—could easily initiate.

In 1982, the same year it was isolated, the bladder cancer gene became the first oncogene to be completely sequenced, i.e., dissected and identified molecule by molecule. Both Weinberg's and Barbacid's research teams sequenced the oncogene and the proto-oncogene from normal bladder cells, then compared the two. The genes differ in several places, but experiments showed that the switch in a single base was the only one needed to give the proto-oncogene the power to transform normal cells in a lab dish into malignant ones. The change puts a thymine unit where a guanine unit should be, and when the cellular machinery "reads" those instructions to make a protein, it substitutes the amino acid valine for glycine at one spot along a chain that is 189 amino acids long. The product of the normal bladder cell proto-oncogene is called p21, but since no one knows what its function is, the function of its abnormal counterpart also remains unknown.

Scientists working with this oncogene believe the abnormal protein itself is somehow responsible for the gene's cancer-causing abilities. If one molecule out of 189 does not seem like a critical change, remember that one swapped amino acid out of 574 makes the difference between normal and sickle hemoglobin. But there is also evidence that increased amounts of the normal protein are enough to transform cells. Douglas Lowy's team at NCI hooked up viral regulatory signals to the bladder proto-oncogene and inserted this hybrid gene into cells in a lab dish. The increased output of the normal product was enough to give the gene transforming powers. This has led some researchers to speculate that there may be more than one way to activate any given proto-oncogene in our cells.

5. Another way for a proto-oncogene to go bad is for the chromosome strand it rides on to be broken off and slapped onto the end of another chromosome. This type of chromosomal abnormality is called "translocation," and has long been associated with certain kinds of cancer, especially cancers of the blood cells.

In 1970, before the tools of genetic engineering became available, researchers had learned to remove DNA from cells, and treat and stain it so that the chromosomes show up as hundreds of dark and light bands. Comparing the bands from normal cells with those from cancer cells, scientists have consistently found specific chromosome defects in most types of cancer: missing bands, translocations, and in a few cases, an extra chromosome.

Jorge Yunis of the University of Minnesota, who has developed high-resolution techniques that reveal thousands rather than hundreds of bands per chromosome set, has predicted that chromosome defects will eventually be found in virtually all cancers. In 99 percent of the cases, he believes, these defects are caused by mutations that happen during a person's lifetime. The other 1 percent are probably inherited.

Despite such findings, researchers have had no idea how and why chromosomes rearrange and what role these abnormalities might play in initiating or promoting cancer. As soon as cancer genes became available, however, it was a logical step to determine if any of them rode on any of the chromosome fragments known to be swapped about or deleted in the cells of actual cancer patients. The answer came quickly: They do.

First, a number of groups mapped the proto-oncogene *c-myc* from normal cells to a spot on chromosome 8. Then they mapped the *myc* gene from patients with Burkitt's lymphoma and found it on chromosome 14. This is the same chromosome swap—between chromosomes 8 and 14—that has been observed in the disease for years, and the results showed that an oncogene is caught up in the shift. But what happens in the translocation to cause the *c-myc* gene to go bad?

Carlo Croce of the University of Pennsylvania's Wistar Institute in Philadelphia believes the shuffling causes the *c-myc* gene to land in a spot where it gets turned on when it should be off. Burkitt's lymphoma is a disease of B cells, the white blood cells that produce antibodies targeted to bacteria and other foreign invaders. In an effort to see what else was in the *myc* gene's new neighborhood after the swap, Croce tracked the location of genes that code for various pieces of antibody molecules, genes that should be some of the most active stretches of DNA in B cells. He found these genes on chromosomes 14, 22, and 2—the same areas where the stray pieces of chromosome 8 carrying the *myc* genes land.

Antibody genes are known to be "jumping genes." During the normal development of B cells, genes for various parts of antibody molecules are assembled from smaller genetic fragments—a process that allows the body to produce an almost endless variety of these defensive proteins. The translocations involving the *myc* gene may occur while this normal splitting, shuffling, and splicing of DNA is going on. The *myc* gene ends up sitting at a site where the natural chromosome breaks occur, next to some of the most active genes in the cell.

Meanwhile, other proto-oncogenes are being mapped to chromosome regions known to be rearranged in the cells of cancer patients. Gallo and Harper, for instance, pinpointed a gene called *fes* from normal cells to the end of chromosome 15, the region involved in a swap with chromosome 17 in a large number of patients with acute promyelocytic leukemia. They mapped a proto-oncogene called *myb* to a region on chromosome 6 that appears to be lost or deleted in a significant number of patients with acute lymphocytic leukemias and malignant lymphomas, as well as in some cases of ovarian cancer and in melanoma, a virulent skin cancer. William Hayward and his group at the Memorial Sloan-Kettering Cancer Center in New York City mapped a gene called *mos* to chromosome 8, some distance away from the *myc* gene. The *mos* gene sits at a spot on the chromosome where breaks have been seen in patients with acute nonlymphocytic leukemia.

The next step is to map the locations of such genes in cells from cancer patients to determine if and where they have moved. Even if these other oncogenes, like *myc*, have jumped, the role of this movement in cancer remains a mystery. Croce's hypothesis about how *myc* gets its malignant powers is still based only on circumstantial evidence, and other researchers disagree. The same questions asked about the bladder cancer oncogene remain for *myc* and others: Does *c-myc* become an oncogene because it suddenly becomes active and starts producing protein when it shouldn't be? Or is it because *c-myc* gets bruised and battered in the chromosome shuffle and ends up making an abnormal protein?

The answers may come as scientists learn more about what the products of proto-oncogenes and their related oncogenes actually do in normal and cancerous cells. This work is just beginning.

In 1983 a group of scientists from four institutions reported finding for the first time a striking similarity between a protein produced by a viral oncogene (*sis*) and a wound-healing protein found in normal human blood. The human protein is called platelet-derived growth factor. After an injury, this protein begins the healing process by triggering cells around the wound to begin multiplying.

The researchers suggested that other oncogenes may also code for proteins that normally stimulate cell growth. "Therefore, any proteins that are suspected in regulation of normal cell growth and all oncogenes should be scrutinized rigorously by computer analysis as soon as their sequences are known," they wrote.

Information about exactly what oncogenes do in cells is crucial to understanding where oncogenes fit into the chain of events that leads to cancer—a process that can often take many years.

Experiments reported by three independent groups of researchers in mid-1983 shed some light on the steps in the cancer process and answered some of the criticisms raised earlier by skeptics such as A. Harry Rubin. He and a few others had cautioned that the transforming powers of the human oncogenes had only been tested in cells grown for years in laboratory glassware, an immortal existence that tends to turn them a little "funny" and might make them susceptible to malignant transformation by certain genes. No one had shown then that oncogenes were capable of turning ordinary cells from a normal person or animal cancerous.

When research teams finally succeeded in inserting oncogenes into normal cells recently removed from rat embryos, they found that it took two events to turn the cells cancerous. The *ras* gene alone only turned these cells slightly abnormal, taking them to the immortal stage needed for life in a culture dish. A second insult, such as insertion of a *myc* gene, was needed to bring on malignancy.

If two or more separate events are needed to cause cancer, this might explain why it takes so long for tumors to develop. At this point, however, the researchers don't know what functions within the cell are being affected by the insertion of oncogenes or the delivery of other carcinogenic insults.

It is already possible to develop simple biochemical assays to screen the cells of both cancer patients and healthy individuals for the presence of the known oncogenes and their products. Some labs are trying to develop highly specific antibodies targeted to cancer genes and their protein products—antibodies that could be used for detection or eventually for therapy. But until the role of oncogenes in cancer is fully understood, the screening results will be of little use in predicting the fate of an individual or devising strategies to interfere with the cancer process.

Other types of genetic analysis, however, are already changing the way doctors approach the treatment of certain cancers, allowing them to individualize therapy for each patient. More than thirty chromosome abnormalities have been linked to various cancers, and Jorge Yunis predicts chromosome defects will eventually be found for most or all of the two hundred types of cancer. The identification of such defects is allowing researchers to classify groups of disorders like leukemias and lymphomas into precise subtypes.

At major medical centers throughout the United States, cancer specialists are already using these findings to analyze the chromosomes of patients for clues to which drugs will be most effective in treating their cancers, whether radiation therapy will help, how fast a tumor is likely to grow, and whether it is likely to spread to other parts of the body. Chromosome defects are also serving as signals to let physicians monitor when a patient's disease is in remission and when a relapse is imminent.

In chronic myelogenous leukémia, for instance, the swap with chromosome 9 leaves chromosome 22 abnormally short. This form of 22 was named the Philadelphia chromosome when it was discovered in 1961 in the cells of a patient hospitalized in that city with the disease. (It wasn't until later that researchers found that the missing piece of chromosome 22 was actually stuck onto chromosome 9.) When white blood cells carrying the Philadelphia chromosome disappear, this signals a patient's leukemia is in remission. If the disease is about to recur, cells with this abnormal chromosome will reappear in the blood well before the patient begins to show clinical signs of relapse.

Many cases of acute lymphoblastic leukemia can be treated successfully today, but when a patient's leukemic cells show a translocation between chromosomes 4 and 11, the prognosis is poor. Chromosome studies can reveal preleukemic conditions among patients with long-standing anemias or abnormal white blood cell counts before clinical symptoms of leukemia develop.

Researchers have known for decades that predisposition to a number of cancers is hereditary. Statistical studies and medical detective work have been used to pinpoint population groups and families with unusually high rates of breast cancer, kidney tumors, leukemias, and other types of malignancy. When a group's life-style, occupations, and other environmental factors do not explain the excess cancer incidence, researchers begin to suspect genetic factors are involved.

But knowing that some unidentifiable "bad" genes are at work is not very helpful to potential cancer victims. Until recent years individuals within identified high-risk populations have had no way to determine their personal chances of developing cancer. For example, women in families with a high incidence of breast cancer can only increase their vigil and hope to spot the disease early if it does develop. Now, however, an intense search is on for identifying markers such as chromosome abnormalities, bio-chemical variations, and even oncogenes that will allow the prediction of cancer risks on an individual basis.

In a few known cancers, such as retinoblastoma, a childhood eye tumor, and Wilms' tumor, a childhood kidney tumor, individuals may inherit chromosome defects that give them a head start toward the disease. Studies of Wilms' victims by Yunis and other researchers show that a piece of chromosome 11 is missing in these individuals' tumor cells as well as in other body cells, indicating that the DNA damage was inherited and present in the fertilized egg. In patients with nonhereditary Wilms' tumor, only the tumor cells have been found to carry abnormal chromosomes. This indicates that a mutation occurred in one or a few cells during the individual's lifetime.

Patients with the hereditary form of retinoblastoma—four out of ten victims of the disease—usually have multiple tumors in both eyes. Their tumor cells as well as their normal blood cells are both missing part of a band on chromosome 13. The same chromosome deletion shows up in the tumor cells of patients with non-hereditary retinoblastoma, but not in their other body cells. The genes involved in both of these inherited childhood cancers, and the event or events necessary to trigger malignancy, are still unknown.

If, as many researchers suspect, DNA must be "hit" twice to cause cancer, perhaps patients with hereditary retinoblastoma or Wilms' tumor are born with one hit. Then when a virus or a carcinogen or an X-ray delivers a second hit or mutation at the same site, cancer is initiated. According to this hypothesis, people without inherited susceptibilities would succumb to the disease only when environmental agents happened to zap a chromosome twice at a crucial site.

In addition to hereditary chromosome defects, several hundred inherited abnormalities in single genes also predispose some individuals to cancer, apparently by leaving them more sensitive to environmental insults and less able to repair DNA damage than most people. Individuals who inherit xeroderma pigmentosum (XP) lack the enzymes needed to repair damage to their DNA caused by the ultraviolet radiation of the sun. Most XP victims get skin cancers before the age of twenty, sometimes dozens or hundreds of them, and many die from the multiple cancers before they reach adulthood.

Ataxia-telangiectasia (A-T) compromises the immune system and the DNA-repair mechanisms and leaves patients' chromosomes fragile and easily broken. Individuals who inherit A-T are extremely susceptible to leukemias, lymphomas, and other types of cancer. Carriers—those who inherit only one gene for the recessive disease—are also more prone than the rest of the population to develop a variety of cancers, from ovarian and stomach cancers to leukemias and lymphomas. About 1 percent of the United States' population is thought to carry a single gene for A-T.

The specific defect in neurofibromatosis is unknown, but patients suffer dozens of lumpy nonmalignant tumors under the skin that can turn malignant. They are also vulnerable to a number of other more serious types of internal cancer.

Hereditary Fanconi's anemia often turns into acute myelogenous leukemia, and some studies indicate that carriers of the recessive gene, like A-T carriers, are predisposed to breast, bladder, lung, and other cancers.

For cancers in which inherited chromosome abnormalities have not been detected, researchers are looking for other types of markers that reveal inherited vulnerability. This often takes the form of HLA types or other genetic characteristics that seem to get passed along in tandem with the unknown gene or genes involved in certain cancers.

One possible marker for breast cancer within breast-cancer-prone families is the enzyme glutamate-pyruvate transaminase, or GPT, which occurs in a variety of forms. The type of GPT a person makes depends on which gene he or she inherits. Mary-Claire King of the University of California at Berkeley studied twenty-one large families that have suffered a high incidence of breast cancer and found that in fourteen of the families, most of the victims had inherited one particular type of GPT. The GPT gene is known to be on chromosome 10, so researchers believe some other nearby gene on 10 must be responsible for a predisposition to breast cancer.

Melanoma, a deadly skin cancer, is on the increase in the United States and elsewhere, and the highest melanoma death rate in this country is in Alabama. Ronald T. Acton's group at the University of Alabama studied 172 melanoma patients and concluded that individuals with the HLA type DR4 seem to be at a high risk of developing the disease. On the other hand, the HLA type DR3 "most likely exerts protective influence on the onset of the disease as well as lessening the severity of the disease should it occur in an individual possessing this (HLA type)."

Other HLA types have been associated with increased risk of cancers of the nose and throat, liver, testes, cervix, kidney, bladder, and rectum, and with Hodgkin's disease. How many of these associations will hold up as HLA types are subdivided and refined remains to be seen.

Most carcinogens in the environment actually have no ability to damage our DNA or initiate cancer until they are taken into the body and activated by normal metabolic processes in the cells. Since genetic factors strongly influence metabolic enzyme processes, all of us vary in our ability to activate carcinogens and, thus, in our susceptibility to cancer. At NCI, Curtis C. Harris and his team are exploring ways to measure an individual's biochemical and molecular responses to cancer-causing agents in the environment and correlate this with cancer risk.

Benzopyrene, one of the most ubiquitous cancer-causing agents, is a component of cigarette smoke, soot, and polluted urban air. Harris's team took cells from various groups of people and found individuals differ markedly—by 50- to 150-fold—in their ability to activate carcinogens like benzopyrene and in the amount of DNA damage they sustain. And people sensitive to one carcinogen, such as benzopyrene, may be quite resistant to the actions of another, like aflatoxin B1, a food contaminant linked to liver cancer.

We all know people who chain-smoked for decades or worked in an industry rife with carcinogens and yet died of "old age" without a trace of malignancy. Tests like the ones Harris and other researchers are developing may eventually be able to tell us what host factors protect such individuals against environmental insults. And Harris hopes these tests will provide each of us with a personal dosimeter that can tell us twenty or thirty years in advance which carcinogens we are vulnerable to and what cancer risks we face. Knowledge of our personal genetic makeup may then help us shape our lives, careers, and personal habits to minimize the risks.

9. Limited Options

... if this thing is ever possible technologically, it will happen. It is no use expecting international agreements to stop it, or a self-denying ordinance among scientists. In fact, most of us, if the power were given us, would be morally confused. For the first application of genetic engineering almost certainly would be to eradicate the grosser genetic misinstructions—that is the prescriptions that produce dystonia or spina bifida or mongoloidism or other fearful forms of human suffering. If you had the power, wouldn't you do that? I would, whatever the consequences. If society had that power and wouldn't use it, I should feel like Ivan Karamazov returning his ticket.

—Lord C. P. Snow, 1973

The islets of Langerhans are no vacation paradise. They are clusters of cells in the pancreas, a large gland tucked behind the stomach. By 1909 medical researchers had figured out that these islets secrete something in normal people that is missing in diabetics. Without this secretion, diabetics are unable to handle sugar properly. In the first decade of the twentieth century, more than 40 percent could expect to die in acute diabetic comas within a few years after their symptoms began. (The missing substance had already been dubbed "insulin" from the Latin word *insula*, island.) But isolating insulin had proved a lot more difficult than naming it.

The islets share the pancreas with other types of cells that secrete digestive enzymes for use in the small intestine. When researchers had crushed up animal pancreatic glands in an attempt to extract the elusive insulin, the digestive enzymes had always destroyed the substance before they could get to it. Then in 1920 a young Canadian physician, Frederick G. Banting, figured out a possible solution, working with a dog: Tie off the ducts between the pancreas and the intestines in the animal, and let the digestive enzymes back up and kill the cells that produced them. After a few months, only the islets should still be functioning. The scheme worked.

In 1921 Frederick Banting and Charles H. Best took an extract from a wasted pancreas produced by their experiment and injected it into a diabetic dog. Within a few hours, the dog's abnormally high blood-sugar level had dropped by half. On July 30, 1921, the two injected insulin for the first time into a human patient, an adolescent boy emaciated by diabetes. The improvement in his health was dramatic, and diabetics began flocking to Toronto for the insulin treatment. By 1923 purified insulin extracted from the pancreases of pigs and cattle became commercially available.

The discovery changed the lives of millions of individuals suffering from juvenile-onset or insulin-dependent diabetes. Today with their blood-sugar levels held in check by daily injections of insulin, relatively few diabetics face the prospect of dying in an acid-induced coma. Insulin therapy has become a prototype for the successful treatment of diseases caused by missing or defective proteins, and sixty years later its impact remains unmatched.

But the picture is not all rosy. If insulin represents one of our best successes in treating inherited defects, it also illustrates how limited our progress has been. Insulin is no cure. Even with one to three injections a day, a patient's blood-sugar level can fluctuate wildly, especially after large meals. Heart disease and stroke are twice as frequent in diabetics as in nondiabetics; blindness is twenty-five times as common, gangrene forty times, and kidney disease seventeen times. More than one quarter of diabetic patients are hospitalized each year, and they have twice as much disability as nondiabetics. By itself, diabetes is still the fifth leading cause of death in the United States, and if complications are added, it jumps to third behind heart disease and cancer. A diabetic's life expectancy is one third less than that of the rest of the population.

Our ability to treat genetic diseases and prevent the damaging expression of inborn flaws has not kept pace with our burgeoning ability to identify and categorize these afflictions and even describe them in molecular detail. In a relative handful of inherited diseases, we have learned to prevent some of the most damaging consequences, soothe the crises, and prolong life. Available treatment approaches range from replacing the missing or faulty gene products to flushing toxic wastes from the body with drugs to altering patients' life-styles or diets so as to avoid exposure to sunlight, barbiturates, or certain foods their bodies cannot deal with. Surgeons can repair congenital abnormalities such as cleft lip and palate or heart defects. Since 1981 researchers have even taken surgery into the womb itself, draining dangerous fluid buildups in the kidneys and brains of the unborn.

Essentially the same strategies are used to treat ailments not traditionally classified as genetic diseases: psychoactive drugs for controlling schizophrenia; diet and life-style changes to prevent heart disease and drugs to steady the heart's faulty rhythms; and surgery, radiation, and drugs to kill cancerous tissue. However, only rarely do we know enough to strike at the roots of such disorders.

The closest thing we have to a cure—the complete correction of a genetic defect at the source—is bone marrow transplants. In a limited number of patients with inherited blood diseases or immune system deficiencies, physicians have been able to effect a cure by replacing their bone marrow cells—and of course the abnormal genes they carry—with foreign bone marrow carrying normal genes. In spirit, the procedure is only one step removed from transplanting the naked genes themselves.

Gene Product Replacement

A number of inherited diseases can now be managed by the same approach used in diabetes—supplying the patient with protein products his own body cannot make because the genetic blueprint for their construction is faulty. Hemophiliacs and patients with other congenital bleeding disorders are treated with injections of clotting factors extracted from human plasma. Children born with pituitary dwarfism can grow to normal heights if they are given growth hormone extracted from human pituitaries. Injections of immunological proteins called gamma globulins can protect children with agammaglobulinemia against life-threatening infections.

A similar approach is to provide a patient with crucial products his or her body cannot make or use efficiently or transport to the right location because an unrelated gene makes a faulty enzyme down the line that blocks the process. In juvenile pernicious anemia, a missing gene product prevents the intestine from absorbing needed vitamin B12 from foods. Physicians can skirt this obstruction by supplying intravenous doses of normal amounts of the vitamin directly into the bloodstream. Larger-than-normal doses of certain vitamins are given to patients with two dozen other genetic disorders to make up for the inefficient ways their bodies process the vitamins to make other needed biochemicals.

In adrenogenital syndrome, production of the steroid hormone cortisol by the adrenal gland is blocked. Without cortisol to provide feedback control, the adrenal glands overproduce male hormones, causing the bones to grow and mature too quickly and fuse early. The result is a too tall child who develops into an unusually short adult. Girls with the syndrome are masculinized. Boys face precocious puberty. Carefully managed daily doses of steroid hormones can damp these effects. As with insulin, clotting factors, and vitamin supplements, this steroid replacement therapy is lifelong.

Costs for many of these therapies are high. The supply of growth hormone extracted from human pituitaries at autopsies is limited. Extracting clotting factors from human plasma can cost as much as $20,000 a year for each of the estimated twenty thousand hemophiliacs in the United States who need them. And about 10 percent of hemophiliacs treated with clotting factors develop antibodies to the foreign proteins or face other problems such as hepatitis because of impurities in the extracts. About 5 percent of diabetics also experience allergic reactions or make antibodies against animal insulins, which vary a tiny bit in their chemistry from human insulin (only one amino acid in insulin from pigs and three amino acids in that from cattle).

Advances in this type of replacement therapy do not have to wait for human gene transplants to be perfected. Genetic engineering techniques applied to microbes are already making improved protein products available for use in patients. Humulin—human insulin churned out by flasks of bacteria which have had human insulin genes installed in them—went on the market in the United States and the United Kingdom in late 1982. Humulin is only the first of a flood of gene-splicing products expected to come up for U.S. Food and Drug Administration approval and to reach the marketplace in the next few years. Many of them will prove useful in the treatment of genetic disorders. Human growth-hormone genes have been inserted into microbes and farmed. Hundreds of labs are racing to do the same with clotting factors and other medically useful hormones and enzymes.

Advances are also being made in the timing and methods of delivering these replacement products to patients. In nondiabetics the pancreas constantly adjusts its output of insulin to keep blood-sugar levels in control, reacting to everything from a cup of coffee with sugar to a full meal with cheesecake. The periodic insulin injections a diabetic takes are timed to coincide only roughly with the highest blood-sugar levels, and many scientists believe this is the basis for most of the complications diabetics face. Since the 1960s researchers have been working to develop various types of artificial pancreases to provide continuous infusions of insulin. In the United States several hundred diabetics are already wearing battery-operated insulin pumps that trickle the hormone continuously through a needle inserted into a vein. In 1980 a Minneapolis man was fitted with the world's first implanted insulin pump. Newer pumps with sensors that measure blood-sugar levels and adjust the insulin flow automatically are in the works. Since many diabetic complications such as blindness and heart disease take decades to develop, it will obviously be a long time before a verdict is reached about the value of these pumps.

Despite such advances, most inherited protein deficiencies cannot be treated yet with replacement therapy. This is because it is difficult or impossible to deliver most gene products to the sites within specific cells where they are needed just by injecting them into the bloodstream. Clotting factors are supposed to circulate in the blood, and insulin and gamma globulin are normally secreted into the blood for delivery to their worksites. But these are exceptions. Most enzymatic proteins are manufactured in the cells where they are needed and never get secreted into or retrieved from the blood, where they would quickly be broken down and lost.

Some modest successes with replacement of circulating enzymes have been reported since the early 1970s, however, and medical geneticists are seeking ways to improve the delivery of enzymes to target cells and slow the rate at which they are broken down and excreted from the body. A human enzyme called glucocerebrosidase, for instance, has been infused into the veins of several patients with Gaucher's disease. Gaucher's, like the more widely known Tay-Sachs disease, is one of about thirty identified "storage" disorders, each of which is caused by lack of a specific enzyme that would normally help cells break down various fatty substances called lipids. Without the enzyme, lipids build up, and in Gaucher's, the lipid is called cerebroside. An accumulation of cerebroside can cause the nervous system to degenerate. Infusions of the missing enzyme reportedly did cause a modest lowering of this lipid level in some patients' tissues. Similar temporary effects were claimed in patients suffering from Fabry's disease, another storage disorder, by giving them intravenous infusions of the missing enzyme, ceramide trihexosidase, or infusions of normal human blood plasma containing the enzyme.

But storage diseases are progressive, and the missing enzyme is needed continuously, throughout life, in all the right cells to halt the devastating effects of the disorders. Various strategies are being devised and tested for trapping enzymes in so-called vehicles—empty shells or "ghosts" of red blood cells or microscopic capsules—to protect them from degradation in the blood and prevent the patient's body from launching defensive antibodies against them. The hope is that enzymes delivered in this way might need replenishment only every few months.

Drug Treatments

Drugs that interact with faulty gene products, or clean up after them, can prevent some of the damaging consequences of a few inherited disorders. Individuals with Wilson's disease cannot process the copper they take in when they eat foods like shrimp, nuts, and chocolate. When the metal builds up to toxic levels, the symptoms can mimic alcoholism or schizophrenia. If the disorder is misdiagnosed and left untreated, the copper poisoning can lead to permanent brain damage and cirrhosis of the liver. A drug called D-penicillamine (a nonantibiotic relative of penicillin), however, can have dramatic effects. The drug grabs onto copper and promotes its excretion from the body. If the drug therapy is started before too much brain and liver damage has occurred, the symptoms are completely reversible. D-penicillamine also helps dissolve the stones that form in the urinary tract of patients with inherited cystinuria, removing a major complication of that disease.

The high blood cholesterol levels of patients with familial hypercholesterolemia can be lowered by drugs such as cholestyramine, which binds bile acids in the gut and causes the body to excrete more of them than normal. This in turn forces the body to convert more cholesterol to bile. The usefulness of such therapy is limited, however, because the liver eventually responds by stepping up cholesterol production. So far, drugs have not proved useful in removing accumulated lipids from the tissues of patients with storage diseases like Gaucher's.

Gout—a disease in which excess uric acid damages the joints and kidneys—can be treated with allopurinol, which inhibits uric acid formation, or with drugs like probenecid, which enhance excretion of uric acid. Lesch-Nyhan syndrome is a much more severe disease than gout, but both involve defects in the same gene. Allopurinol is also effective in reducing the uric acid concentrations in Lesch-Nyhan patients, preventing the kidney damage that used to kill most victims of the disease in childhood. Patients in their twenties are alive today. Unfortunately, however, the treatment has no impact on the most devastating symptoms of the disorder—the mental retardation, cerebral palsy, and compulsive biting and self-mutilation.

A limitation on wider use of this type of therapy is that it requires drugs that act on specific enzymes or other substances, grabbing or blocking them as needed. Specifications for potentially useful new drugs are being worked out by medical researchers using the new tools of molecular biology to probe the underlying defects in a number of genetic disorders.

Sickle cell anemia is one such disease. Scientists have a thorough understanding of the molecular, even atomic interactions that cause the blockage of blood flow and bring on the painful sickle cell crises that can damage the spleen and compromise the immune system. Until recent years patients have had to face these crises with only pain-killers, antibiotics, and plasma and water to increase their blood volume, and in the most severe cases, with blood transfusions. As we saw earlier, sickle cell patients' red blood cells distort into rigid sickle shapes only when oxygen levels are low, after the cells have delivered the oxygen they carry to the tissues. When the cells reach the lungs and take on a new load of oxygen, they become round and flexible again.

William A. Eaton and James Hofrichter at the National Institute of Arthritis, Metabolism, and Digestive Diseases in Bethesda, Maryland, have described the conditions that apparently bring on a crisis: Normally by the time they sickle, most of the red blood cells will have gone through the tiny capillaries where they release their oxygen and be in the veins on their way back to the lungs. But under certain conditions, including the high temperature of a fever, the cells may release their oxygen sooner than usual and distort while they are still in the capillaries. The rigid crescent cells block the tiny blood vessels, cutting off the supply of oxygen and damaging the surrounding tissues.

With this understanding, a number of research groups are developing and testing drugs designed to prevent the sickling, or to delay it so that the cells have time to reach the veins where they are less likely to cause severe damage.

A less well understood genetic disease is Duchenne muscular dystrophy. Scientists are still working to pinpoint the defective protein responsible for the muscle wasting that the disease's victims suffer, and the defective gene behind it. But even without that understanding, researchers have gained some insights into the disease process itself in the past few decades and have come up with treatment strategies that might control its progress. Research suggests that the culprit is large amounts of calcium that leak from the bloodstream through the outer membranes of muscle cells and, once inside, trigger the deterioration of the muscle tissue. This finding has led scientists in several countries to launch tests in patients using drugs that prevent calcium from getting into muscle cells—the same "calcium blockers" that have been developed and approved in the last few years for treating heart disease.

Diet and Environment

In an increasing number of inherited disorders, researchers are finding that some environmental trigger is needed to bring on the severest consequences. By avoiding specific environmental factors that their genes have left them unable to handle—drugs, foods, sunshine, cigarettes—individuals can often lead fairly normal lives. The classic example of such a disorder is phenylketonuria, or PKU, a success story as heartening to geneticists as the discovery of insulin.

In 1934 Norwegian physician Ivar A. Folling examined two mentally retarded siblings and found something different about their body chemistry: Their urine turned a rich green when ferric chloride was added to it. Normal urine does not turn green under these conditions. The urine from these children turned out to con-

tain large amounts of something not normally found in urine, phenylpyruvic acid. The condition was named PKU and discovered
to be hereditary. Today we know that an abnormal recessive gene
in PKU victims is responsible for a deficiency of a single liver
enzyme—phenylalanine hydroxylase. This enzyme is supposed
to break down the amino acid phenylalanine to another amino
acid, tyrosine. When this pathway is blocked, the body tries to
deal with the excess phenylalanine from proteins in the diet by
converting it to phenylpyruvic acid and related chemicals. The
buildup of these chemicals in the body prevents a child's brain
cells from developing normally, and the victim ends up severely
retarded.

In the 1950s researchers working with PKU victims had an
idea: If their bodies are not able to handle excessive phenylalanine, then don't feed it to them. Perhaps if these chemicals
were never given a chance to accumulate, the brain damage could
be avoided. The idea worked. At first physicians worked only
with families that already had one or more children with PKU and
thus were known to be at risk of having another. As soon as a new
baby was born to such a family and found to have high phenylalanine levels, a strict diet was prescribed to make sure the
infant took in only the minimum amount of the amino acid needed
for life and growth. The results were dramatic: The children grew
up with I.Q.s averaging only slightly below mean for the population. In 1961 a simple blood test for detecting PKU was developed
by Buffalo physician Robert Guthrie. Today the vast majority of
newborns in the United States, western Europe, England, and
Israel undergo routine PKU screening.

Since then, a number of other inborn errors have been found
that respond to this type of diet-restriction therapy. Patients with
other disorders of amino acid metabolism, such as homocystinuria and maple syrup urine disease, benefit from avoiding
the substances they cannot metabolize—homocystine or meth-
ionine in the former, and valine, leucine, and isoleucine in the
latter. Diets free of milk and milk products can prevent severe
diarrhea in individuals with lactase deficiency, and the potential
for starvation or severe liver disease, cataracts, and mental retardation in children born with galactosemia. Individuals with familial hypercholesterolemia can reduce their risk of heart disease by
eating less cholesterol.

Diet is only one of the factors known to aggravate inborn susceptibilities. A whole list of drugs can trigger hemolytic anemia, the breakdown of red blood cells, in individuals deficient in glucose-6-phosphate-dehydrogenase (G6PD), an enzyme used by red cells in the breakdown of sugar for energy production. These medications include the antimalarial agent primaquine, several vitamin K preparations, some fever-reducing drugs, sulfa drugs, and even aspirin. Exposure to certain chemicals, like naphthalene in the workplace, can also bring on anemia. A related defect in the gene for G6PD causes favism, a severe anemia brought on by eating fava beans, a common crop in Mediterranean countries, or inhaling pollen from the bean plants.

Anaesthesiologists use the muscle-relaxing drug succinylcholine to relax a patient's vocal cords and ease the task of inserting a tube down the windpipe before surgery. However, if the drug is administered to a person with an inborn deficiency in pseudo-cholinesterase, the patient may stop breathing for as long as several hours and must be kept alive on an artificial respirator until the effect wears off. Liquor or barbiturates can make a victim of porphyria behave like a schizophrenic.

Eventually almost all victims of the severest form of alpha-1-antitrypsin deficiency end up with emphysema—if they survive cirrhosis of the liver in childhood; cigarette smoking significantly accelerates the progress of their lung disease. The most effective therapy for prolonging the lives of xeroderma pigmentosum victims, who lack the ability to repair DNA damage caused by ultraviolet light, is to keep them out of the sun.

Occasionally, physicians have been able to begin this type of therapy quite early in a patient's life, manipulating the womb environment itself. In 1975 physicians diagnosed a fetus with a rare inherited disorder that left it unable to synthesize vitamin B_{12}. The defect, left untreated, leads to mental retardation and often death. The team at the Tufts-New England Medical Center in Massachusetts, treated the condition successfully by having the mother take massive doses of vitamin B_{12} during the last months of her pregnancy. In 1980 physicians at the University of California at San Francisco performed amniocentesis on a pregnant woman who earlier had given birth to a child with a life-threatening defect in biotin metabolism (biotin is one of the B vitamins, formerly labeled vitamin H). The test showed the fetus the woman was carrying suffered from the same defect. The mother was given large doses of biotin, and the infant girl was born without symptoms of the deficiency that had almost killed her brother at birth.

Avoidance therapy in the womb is also crucial for a special group of babies to keep them from becoming victims of a genetic disorder most of them do not even inherit. These are the children of the thousands of women around the world with PKU who have been saved from retardation since the early 1960s by special diets. Most stopped the diets as they grew up, when their brain development was complete and the high phenylalanine levels brought on by normal protein intake were no longer a threat to their intelligence. But high levels of this amino acid in a pregnant woman's blood will irreversibly damage the brain of her developing baby. Putting women with PKU back on carefully tailored diets during pregnancy is a relatively new therapy, and the results are still mixed. Women who start the diet before they become pregnant seem to have the highest likelihood of bearing babies with normal intelligence. Some physicians, however, still encourage women with PKU to put off having children until more study has been done.

As researchers identify the precise genetic and protein defects in more and more inherited diseases and vulnerabilities, the opportunities for therapy by life-style or dietary manipulations will undoubtedly increase.

Organ and Bone Marrow Transplants

In the 1960s, when gene transplants still looked like a twenty-first-century fantasy, geneticists conceived of using another relatively new medical procedure as a possible way of installing a permanent supply of "good" genes into patients with genetic defects: tissue and organ transplants. The cells of foreign organs should contain normal genes the patient's own tissues lack and be able in some cases to supply at least a small amount of the missing gene product.

Patients with the storage disorder Fabry's disease seemed to be compelling candidates for organ transplant therapy, and not just as a way of giving them new genes. The fatty lipids that accumulate in the victims' blood vessels commonly lead to progressive kidney deterioration, and those who don't die from strokes in early adulthood usually do so from kidney failure. The first studies in the 1960s involved kidney transplants in three of the Fabry's disease patients, and the results were promising: Following the transplants, levels of the missing enzyme (ceramide trihexosidase) reached about 10 percent of those found in normal individuals. That was apparently enough to do the job. Within six months the continuous supply of the enzyme had removed accumulated lipids from the patients' own kidneys and other organs. Their other symptoms had cleared up and they considered themselves cured. Attempts to repeat these successes in other patients, however, brought less encouraging results.

Geneticists have also dreamed of transplanting kidneys into patients with genetic diseases in which kidney damage is not a problem. Drugs are available to prevent kidney damage from excess uric acid levels in Lesch-Nyhan patients, for example, but some geneticists would like to try transplants for another reason. They envision implanting a tissue that could provide at least a small quantity of the missing enzyme in hopes it could prevent or reverse the symptoms that defy other available treatments. In treating Lesch-Nyhan, the main concern is brain damage, which cannot be prevented by controlling uric acid buildup. But too many obstacles have stood in the way of this type of experimentation. Relatively few donor organs are available, and there are long waiting lists of patients with kidney failure and other life-threatening medical needs which take precedence. Also, a donor organ with a similar tissue type must be used to lessen the chances that a patient's body will reject it as foreign.

Rejection has not been a problem when transplants are done in children born with defects that leave them with little or no immune defenses of their own. The thymus gland lies above the heart, behind the breastbone. Early in childhood, this gland plays a crucial role in the maturation of certain white blood cells called T lymphocytes that form an important part of the body's immune system—cellular immunity. The thymus grows rapidly during the first two years of life, gradually diminishes after puberty, and ends up as a lump of fatty tissue.

The first thymic transplant was performed on an infant with DiGeorge's syndrome who was born missing the parathyroid glands in the neck as well as the thymus. The child had cataracts in both eyes at birth and developed a cluster of chronic infections during the first month of life. Using a thymus taken from a human fetus, surgeons cut the gland into three pieces and implanted them into their tiny patient's abdominal muscle. The results were immediate. His lymphocyte count rose, the infections began to clear up, and his immune system began to develop normally.

The body has another type of defense in addition to the cellular immunity provided by the T lymphocytes. This is the humoral immunity that is provided by circulating antibodies and immunoglobulins, which grab onto bacteria, foreign proteins, and other invaders, and incapacitate them or mark them for destruction. Children born with severe combined immunodeficiency disease lack both types of immunity and usually die from infections in early infancy.

A group of physicians at the University of Minnesota thought that an infusion of normal bone marrow might provide a cure for this affliction. Marrow contains primitive cells called stem cells, which can generate both lymphocytes and the plasma cells that produce antibodies and immunoglobulins. In 1968 the team tried the procedure for the first time on an infant boy born with a combined immunodeficiency.

Although rejection of the transplanted marrow by the child's body was not a worry, a converse danger still made it necessary to find a donor with a compatible tissue type. What was feared was an attack by the foreign marrow against the patient himself. A foreign kidney cannot do this, but immunologically competent marrow cells can respond to the new patient host as though he were the foreign invader. This is called the "graft versus host reaction," and up to that time it had killed all recipients of bone marrow grafts. In this case, one of the boy's sisters turned out to have a compatible tissue type.

It was touch and go for months after the first injection of bone marrow, as the foreign material attacked various components of the boy's own blood and marrow. After his red blood cells had been completely destroyed, the team tried a bold maneuver—a second infusion of foreign marrow. This time, with little of the child's own blood-forming system standing in its way, the foreign marrow took over peacefully and repopulated his system with its own red cells and other blood components. The boy's inherited defect was cured, and he proceeded to grow normally.

Since that 1968 success, bone marrow grafts have been used more widely than any other transplant technique in treating genetic defects. In 1982 physicians at the Fred Hutchinson Cancer Research Center in Seattle reported that they had used bone marrow transplants for the first time to cure two patients with the inherited hemoglobin defect beta-thalassemia. "Marrow transplantation offers the chance to cure not only thalassemia but also other genetic diseases of the marrow, such as sickle cell disease," the medical team wrote in their report to the British journal *Lancet*.

(Marrow transplants in thalassemias, aplastic anemia, and similar disorders in which the patient has an intact immune system require that the individual's own marrow be destroyed by total body irradiation before the transplant.)

Despite these successes, marrow transplants are still available only to a handful of patients, since the majority of those who might benefit do not have a tissue-compatible donor.

But many groups are working to change that situation. Researchers at the Memorial Sloan-Kettering Cancer Center and the Weizmann Institute in Israel are using a soybean product called lectin as a sort of glue to latch onto the mature T lymphocytes and separate them from the rest of the marrow. Eight children with severe combined immunodeficiency were treated in the first trials with the nonmatched but lymphocyte-free marrow. Seven of the children recovered, apparently cured. As new T lymphocytes are produced by the stem cells in the transplanted marrow, they are processed by the host's thymus and thus do not recognize their new environment as foreign. Other groups have used specially tailored antibodies and even sheep red blood cells to rid donor marrow of T lymphocytes before transplant.

Other scientists are trying to find tissues that provoke no immune response and that could someday be engineered to produce enzymes or other gene products when transplanted into virtually any patient. Researchers at Guy's Hospital in London are already taking bits of the amniotic membrane that wraps the fetus and attempting to transplant them into patients to see if they can supply missing enzymes. The amniotic tissue apparently incites no defensive response from the recipient's body. The team has speculated that such cells might be engineered to produce a variety of gene products that patients need, proteins that are not normally produced by amniotic cells. Other research teams are doing similar but more controversial experiments with tissues taken from human fetuses.

Tissue transplantation clearly has a long way to go in order to make a major impact on the treatment of patients with genetic disease, but scientists believe it could eventually be put to wider use. Yale University geneticist Leon E. Rosenberg has suggested that spleen transplants might provide lasting benefits to hemophiliacs and patients with agammaglobulinemia, and liver grafts could supply missing enzymes to individuals with PKU, urea cycle defects, and diseases involving abnormal accumulations of glycogen (stored carbohydrates). Several researchers around the world have tried transplanting whole pancreases or insulin-secreting islet cells into human diabetics, but success has been very limited. Several groups have succeeded in bringing blood-sugar levels under control in diabetic rats or mice by transplanting into them islet cells from healthy rodents. The hope is that animal islets might someday do the same for human diabetics.

Despite this wide assortment of treatment methods and a few heartening success stories, physicians still have little but comfort to offer victims of the vast majority of genetic disorders. Even abortion—which can hardly be called therapy—is an available option only in the small percentage of genetic diseases that can be routinely diagnosed in a fetus before birth. Every year in the United States, for example, two thousand to three thousand children are born with cystic fibrosis. The disease is incurable, and it cannot be detected before birth. The genetic defect involved has not been pinpointed, but somehow it causes mucus secretions in the body to become thick and sticky, clogging the lungs and blocking the pancreatic and bile ducts. Thirty years ago most of these children died at three or four years of age. Now, with physical therapy and special breathing exercises to help clear their airways, cystic fibrosis victims can often live into their twenties. The extended life-span, in this and many other genetic diseases, is not an unmixed blessing for the patient or his family. The daily routine of therapy, medications, and controlled diet, plus the certainty of death at a young age, can create emotional and financial burdens that tear families apart.

As molecular biology techniques allow us to unveil the defects behind cystic fibrosis and other inherited disorders, we will undoubtedly develop more effective therapies to ease their devastating symptoms. In some cases this may be all that is needed. Physicians and patients may never feel the need for more permanent solutions. In certain growth-hormone deficiencies, for example, a limited number of injections during childhood and adolescence are all that are necessary to produce a normal-sized individual. With an abundant supply of human growth hormone produced by genetically modified bacteria becoming available, trying to implant genes for growth hormone into patients might look like heroic overkill. But most conventional treatments are not so effective and convenient. They are lifelong, costly, and— most important—do not alleviate all the harsh effects of a disease.

Gene therapy will not be a panacea. In its first applications, in fact, it will be as limited as bone marrow transplants were a decade ago. It is, however, an inevitable goal in our long struggle to intervene at the roots of genetic disease.

III
PROBING GENES

10. Isolating Genes

Just as our present knowledge and practice of medicine
relies on a sophisticated knowledge of human anatomy,
physiology, and biochemistry, so will dealing with disease
in the future demand a detailed understanding of the
molecular anatomy, physiology, and biochemistry of the
human genome.

—PAUL BERG, 1980 Nobel Prize lecture

Viewed in computer terms, our genetic legacy is a reel of microscopic mag tape three feet long, carrying three billion bits of
information in a four-symbol language—AGCT. More than 95
percent of the tape may be junk and nonsense. But interspersed
along its length are fifty thousand to one hundred thousand crucial
bytes, messages that are each a few hundred to thousands of bits
long, which encode all our physical and chemical characteristics.
Even these bytes—our genes—may be interrupted one to fifty or
more times by stretches of garble. If we could read the whole
tape, the significant messages would at first be indistinguishable from the junk amid the monotonous sequence of TGAG-
CAAATGGGTGATCCAGATA . . .

The development of recombinant DNA techniques in the past
ten years, however, has given us a way to pull single genes from
this welter of tape and decode them bit by bit. Readouts of genetic
sequences from the DNA of bacteria, mice, cattle, people, and
other living things are being collected in computer banks at the
rate of more than a million molecular bits a year, and the pace of
the effort is accelerating constantly. Long before we have deciphered the whole human tape, however, scientists hope to pull
out and decode the small percentage of the sequence that carries
the program or blueprint for making a human being. And once we
know how "normal" human blueprints read, each with its individual load of sublethal quirks and glitches, we should be able to
pinpoint the genetic errors that result in serious disease and
suffering.

Among the first human genes to be isolated and sequenced were the globins, which code for protein chains that make up the hemoglobin molecule. More than three hundred variant forms of hemoglobin have been spotted, many of them leading to anemias and other blood disorders, and researchers have now isolated and decoded a number of these defective genes. Other human genes now in hand include several for collagen, a structural protein used to build skin, bones, and ligaments; insulin and growth hormones; several virus-fighting interferon proteins; enzymes like HGPRT and G6PD, which are needed to run the basic biochemical processes of the body; and many antibody and other immune system components.

To isolate a specific gene, a researcher first has to have some clue to what he is looking for. He needs a "probe," a molecular hook that can identify and latch on to the gene in a crowd. This requirement is less impossible than it might seem, thanks to the nature of the genetic code and the information flow within a cell.

Each gene carries a set of instructions for manufacturing a protein. But the genes are safely locked away in the protective vault of the nucleus, unavailable to the cell's protein factories, the ribosomes. Working copies of the genetic instructions, with all the garbles and interruptions cut out, must therefore be made and sent out to the factories. These copies—called messenger RNAs or mRNAs—are written in a four-symbol code only slightly different from the DNA code (the T for thymine is replaced by a U for the chemical unit uracil). It is these mRNAs that first made gene isolation possible.

Some viruses routinely carry their genetic information in RNA instead of DNA. When certain of these viruses (called retroviruses—more about them in Chapter 22) enter a cell, they have to convert their instructions to the native DNA language of the cell in order to operate there. To do this, these viruses have invented an enzyme called "reverse transcriptase" that can translate an RNA sequence into the equivalent DNA sequence. It is an enzyme that genetic engineers have taken full advantage of in the past decade. A DNA copy of an mRNA which is synthesized by using this enzyme is called complementary DNA or cDNA to distinguish it from the original gene. (The mRNA copy and the cDNA are all message and no interruptions, unlike the gene.)

A researcher who can get his hands on an mRNA can eventually obtain the gene from which it was copied, using either the mRNA itself or a cDNA copy as a probe to search through a crowd of genetic material and fish out the original gene. The procedure, called molecular hybridization, is made possible by the complementary nature of the two twisted strands of the DNA molecule. Remember, an A on one strand is always paired with a T on the other, and G is always paired with C. When the strands are separated, this complementary pairing system allows them to find each other again and lock together with precision like the two halves of a zipper. A sequence of ATGCAC must pair up with TACGTG. In the lab, researchers use heat or alkali to unzip and separate the strands of DNA. Then they place the DNA on the edge of a hard sheet of gel and apply an electric current to make it spread out across the gel. Quantities of an mRNA or a cDNA probe with a radioactive label or tracer attached to it are dumped on the DNA. As the mixture cools again, the DNA begins to pair up, and the single-stranded probes link up with complementary segments of DNA—the original genes from which the probes were copied. The radioactively tagged genes can then be retrieved from the rest of the DNA.

Molecular hybridization techniques were developed in the early 1960s, but they didn't make gene isolation an overnight success. The problem is that many of the most interesting genes, such as interferon, are present in only one or a few copies per cell. A probe that must look for and link up with a few thousand bits among billions has a needle-in-a-haystack task. Even getting a probe may be difficult if the gene is active only occasionally in a small population of cells, and if mRNA copies are usually scarce or absent.

In the late 1960s, before the advent of cloning, the first few genes of known function were isolated from toads and bacteria. But these were special cases. The isolated toad genes were present in thousands of copies per cell and were known to be slightly different in chemistry and density from the rest of the toad DNA. This density difference allowed researchers at the University of Edinburgh to spin the DNA in a centrifuge and separate out a relatively concentrated solution of the desired genes before using the hybridization probes.

In another early procedure, Harvard University biologists got an enriched solution of the bacterial genes they were trying to isolate by first allowing viruses to infect the bacteria and kidnap the genes. (Viruses make a habit of this, as we will see in Chapter

14.) Genes that had made up only one tenth of 1 percent of the bacterial DNA then made up 5 to 10 percent of the DNA in the virus particles. (When the isolation of the bacterial gene was announced at a press conference in 1969, London newspapers reacted with alarm: "Genetic 'bomb' fears grow," the *Evening Standard* proclaimed. "The frightening facts of life. Scientists find secret of human heredity and it scares them," the *Daily Mail* reported.)

But tricks like these can be used only for the isolation of a limited number of genes. The technique that makes isolation of virtually any gene possible is molecular cloning. This process allows researchers to fragment a cell's DNA and mass-produce thousands of copies of each piece, thus giving a probe a better shot at finding the gene it is looking for.

The first step in cloning is making a hybrid or recombinant DNA molecule that will be capable of entering a bacterial cell and reproducing there. Scientists start with a suitable vehicle, either a plasmid, a small circular loop of bacterial DNA, or a small virus called a phage, which infects bacteria. The vehicle must then be cut open and a piece of foreign DNA spliced in. If this sounds like a surgical procedure, it is really more like cooking. Both the chopping and splicing of these microscopic DNA fragments is done in test tubes or flasks by chemical scissors called "restriction enzymes" and glues called "ligases."

Restriction enzymes and ligases are among the many invaluable substances, including reverse transcriptase, pirated from viruses and bacteria since the early 1970s that make modern genetic engineering possible. Bacteria use restriction enzymes to destroy foreign DNA, cutting invading phages to ribbons by chopping the DNA at precisely predetermined points. One enzyme, for instance, cleaves a DNA strand between the G and A every time the sequence GAATTC appears. Various bacteria have developed different enzymes and protective substances to make sure their scissors don't cut their own DNA. Researchers have been able to isolate more than 150 of these chemicals, each with its own cleaving target.

Scientists use restriction enzymes both to chop open the phage or plasmid vehicles and to hack the cellular DNA they want to clone into approximately gene-sized fragments. No elaborate tricks are needed then to join these fragments into recombinant molecules. When certain restriction enzymes cut double-stranded

DNA, they don't slice straight across both chains. The cuts are staggered so that at each end a single A, T, G, or C unit is left dangling without a partner. These are called "sticky" ends. DNA fragments cut with the same enzymes have complementary sticky ends and don't need much encouragement to link up. Ligases complete the job of sealing the bond. (Not all vehicles, of course, will pick up lengths of cellular genes. Some plasmids will stick to plasmids, some cellular DNA fragments to other fragments. The recombining process is random.)

This flask of recombinant molecules is then introduced to bacteria. Some bacteria will take in plasmids (or be infected if phages are used as the vehicles) and treat this foreign DNA like their own, making new copies of it and passing them along to each new generation. Phages and plasmids also multiply independently within their hosts, giving rise to dozens or hundreds of copies of themselves within a single bacterial cell. A weakened and domesticated variety of a common bacteria found in the human gut, *Escherichia coli* (*E. coli*), has become the workhorse of cloning, reproducing itself and the foreign DNA it carries every twenty minutes, like a molecular factory.

Because scientists do not want to house and feed any freeloading bacteria that fail to pick up recombinant molecules, they have devised selection procedures to kill off the nonproducers. One way is to use as a cloning vehicle a plasmid that carries a gene for antibiotic resistance. Normal *E. coli* that don't have this gene are very sensitive to certain antibiotics. After researchers have exposed *E. coli* to recombinant plasmids carrying the resistance gene, they subject the bacteria to antibiotics. Any bacteria that failed to take in the foreign DNA are killed.

When a scientist chops up the entire DNA content of a human or other cell into millions of fragments and introduces them all into bacteria for cloning, the procedure is called "shotgunning." (In the mid-1970s shotgunning worried some scientists and laymen concerned about the safety of recombinant DNA experiments. No one knew what tricks bacteria might learn when offered a full range of other creatures' genes. Researchers have since found, however, that lab strains of *E. coli* used for cloning are apparently too tame to escape and survive on their own even with a load of foreign DNA.)

With probes available, screening each of the bacterial colonies that result from shotgun cloning to find which carries the wanted gene is tedious but manageable. The most difficult part of gene isolation today is getting probes, especially for single-copy genes and genes that are active only at low levels. Two newer approaches promise to expand greatly our ability to fish out such genes. One involves transferring genes into animal cells rather than into bacteria, and the other makes use of man-made probes.

When genes from bacteria or viruses are cloned in *E. coli*, they may betray their presence by getting down to work and producing protein products in their new settings. This makes it unnecessary to have mRNA or cDNA probes to find them. Human genes, on the other hand, don't usually work in bacteria until they have been specially engineered to function there, as biologists have done with insulin, growth hormone, and interferon genes. Some human genes will function in animal cells, however, and by the late 1970s, biologists had learned several ways of transplanting genes into animal cells growing in lab dishes. The first generation of techniques for transferring naked bits of DNA into animal cells are not as efficient as shotgunning DNA into bacteria. But these techniques have allowed the isolation of a number of genes for which researchers had no probes.

Patients with Lesch-Nyhan syndrome, for instance, are missing a crucial enzyme with a four-dollar name—hypoxanthine guanine phosphoribosyl transferase, or HGPRT. Several years ago researchers at the University of California at San Diego (UCSD) set out to isolate the HGPRT gene. Although they had no probe, they did have a selection system that could tell them when the HGPRT gene was present and functioning somewhere in the DNA they were working with. Mutant mouse cells that have no working HGPRT genes of their own are available from biological supply houses. These mouse cells cannot live when their dishes are filled with a nutrient broth containing the chemicals hypoxanthine, aminopterin, and thymidine—dubbed "HAT" medium. Only cells that make the enzyme HGPRT can survive in HAT.

Using the DNA transformation techniques described in Chapter 16, the UCSD team led by Theodore Friedmann exposed these mouse cells to DNA taken from normal human placental cells. A small fraction of the mouse cells absorbed bits of the human DNA just as bacteria take in plasmids, and these DNA fragments sometimes included an HGPRT gene. When the mouse cells were placed into HAT medium to grow, only the ones that could put their new HGPRT genes to work survived and grew. The researchers knew the human HGPRT gene was present in the surviving cells, but so were other bits of human DNA. To try to eliminate this other DNA, the team extracted the DNA from this first batch of surviving mouse cells and offered it to another batch of deficient cells. After this second cycle, most of the human DNA that had no survival value for the mouse cells was eliminated, and the surviving cells carried only functioning copies of the HGPRT gene.

The next problem was to retrieve the human gene from the mouse DNA without the benefit of an HGPRT probe. Since the researchers could not spot the gene directly, they took another tack. They looked for one of the bits of garble (called an "Alu repeat sequence") that sprinkle the human DNA strands, hoping that one of these sequences would be sitting nearby or even interrupting the HGPRT gene. Sure enough, using a radioactively labeled Alu sequence as a probe, Friedmann's team was able to grab a fragment of the human HGPRT gene from the mouse material. Finally, the researchers were able to use this gene fragment as a probe to fish out a whole HGPRT gene from chopped-up human DNA. (A team at Baylor College of Medicine in Houston fished out the human HGPRT gene using a cDNA probe made from copies of the mouse HGPRT gene.)

Other labs are using similar procedures to isolate human cancer genes. When DNA from human tumor cells is transferred into certain mouse cells, some of the cells turn malignant, indicating they have picked up a human gene with cancer-causing potential. The unknown human genes responsible for this change can then be isolated from the mouse DNA using this method.

Such techniques are still of limited use in isolating genes. Few human genes have been made to function in foreign settings under their own steam, whether the cells are mouse, human, or bacterial. Also, as mentioned earlier, the first generation of gene transplant techniques is inefficient. Only a fraction of cells will take in new genes, and without a selection scheme like HAT, or a visible change like malignancy caused by a foreign gene, there is no way for scientists to identify whether any cells have taken in working copies of the gene they are interested in.

New gene transplant methods that could get every fragment of a cell's DNA into foreign cells and make each piece function would open a floodgate of opportunities for gene isolation. A second generation of techniques already developed may provide just that.

"Initially, people got the easy genes, like the globin genes, the immunoglobulin genes, the genes for which you have probes, for which you can easily isolate rich messenger RNA preparations that would make cDNA," says Richard Mulligan of MIT, whose team is trying to domesticate viruses to carry virtually any gene into a foreign cell and force it to work there. "But in the future, particularly in the commercial biotechnology companies, one would like to isolate genes for very potent biological products that may be expressed in only a small percentage of cells, or expressed at very low levels in all cells. And right now, if you don't have a probe, you can't isolate them."

Mulligan envisions starting the gene isolation process using cDNA "libraries" instead of the total DNA of a cell, which is largely junk and garble. These libraries are made by reverse-transcribing all the mRNA extracted from a cell and thus ending up with cDNA copies of all the genes that are actively making protein in that cell. These cDNAs have none of the control signals needed to make the natural gene function, but Mulligan's viral vehicles could supply the signals. Each of the cDNAs could be inserted into a foreign cell and turned on; researchers could monitor the cells' protein products to see what each gene produced. The cDNA then could be retrieved from each cell and used as a probe to grab the natural gene.

"We've actually demonstrated the power of the technique, and it does work," Mulligan says. "You can find things that are very rare . . . And Paul Berg [of Stanford] has worked out a very efficient way of making these cDNA libraries." Scientists thus may be able to isolate genes that code for a wide range of novel enzymes and proteins even before they know what these products do in the human cell.

Other new techniques are providing scientists with synthetic probes to use in isolating genes. A researcher who has isolated a tiny dab of a protein that looks medically useful or seems to be involved in a genetic disease can now build a probe for the gene without having to isolate the mRNA. This technique makes use of automated "gene machines" that can assemble artificial DNA strands according to recipes devised from analyzing the protein.

In the protein factories of a cell, the four-symbol code of an mRNA is translated into a twenty-symbol code. Each symbol designates one of the amino acids from which protein chains are assembled. Sequences of three units from the DNA code translate into one amino acid. The triplet GCU, for instance, codes for alanine. Thus the type and order of amino acids in every protein reflect the molecular sequence of the gene that codes for that protein. Translating backward is not foolproof, because there are cases in which an amino acid is coded for by more than one DNA triplet. Tyrosine, for example, is coded for by either UAU or UAC. But anyone with a list of the codes can quickly draw out all the possible translations—on paper or by computer.

This knowledge cannot be applied directly to gene isolation, however, because a protein and its parent DNA do not recognize each other and will not hybridize or bind together the way mRNA and DNA will. Proteins therefore are not suitable probes. But techniques are now available to construct synthetic DNA probes, by hand or by machine, once the amino acid sequence of a protein is known.

Starting in 1945, British biochemist Frederick Sanger developed a method for breaking the amino acids from a protein chain and identifying their order. (He announced the structure of the protein insulin in 1953, the same year Francis Crick and James Watson announced the structure of DNA.) By 1967 the process had been automated. An instrument called a "sequenator" could peel the amino acids off a protein and identify them one by one. But the first sequencing machines required five thousandths of a gram of pure protein, and many proteins, especially critical enzymes, hormones, and brain messenger chemicals, cannot be isolated in those quantities.

Today at the California Institute of Technology (Caltech), biologists Leroy Hood, Michael Hunkapiller, and William Dreyer are testing an instrument called a "microsequenator" that can snip amino acids from protein samples as small as a few ten millionths of a gram. A second instrument identifies each amino acid as it is removed. The biologists are working with engineers at NASA's Jet Propulsion Laboratory at Caltech to replace this second instrument with an ultrasensitive mass spectrometer that will make the process ten times more sensitive.

With the sequence of the protein—or even part of it—in hand, researchers can quickly translate it on paper or feed it into a computer and work out all the possible combinations of chemical units for the gene that ordered it. The next step is to assemble a synthetic strand of DNA, twelve to twenty chemical bases long, that corresponds to a stretch of the natural gene (or if several translations of the protein are possible, several synthetic DNA strands). This will be the probe used to fish for the natural gene.

During the 1950s researchers began learning to string molecules together like beads to form artificial DNA sequences. By 1976 H. G. Khorana of MIT had assembled an artificial gene 199 units long that actually functioned when it was inserted into a bacteria. Today the Caltech group and several other university and commercial research teams have developed DNA synthesizers or gene machines to automate and speed up the task. A researcher simply punches the code into a keyboard, and a microprocessor takes charge, systematically releasing reagents, solvents, and other chemicals onto a small disk where an artificial gene is assembled bit by bit. This synthetic gene fragment can be introduced into bacteria, and it can be mass-produced by cloning just like natural genes.

Isolated genes today number in the dozens. With the available techniques, Hood predicts the number will expand into the hundreds in the next five years. Many of these genes will be ones that cause genetic diseases when their sequences get jumbled. With the genes in hand, physicians will be able to screen fetuses for a much broader range of inherited disorders, and researchers can begin devising therapies to curb the damaging influence of defective genes.

For people who worry that genetic engineers will attempt to tamper with human nature as well as with disease, the realities of gene isolation should illustrate how difficult it is going to be to find genes that influence complex traits like intelligence, temperament, and character. Intelligence cannot be spotted in a dish of cultured cells. Trying to find an "intelligence gene" by transferring thousands of unknown human genes into cells and examining all the novel proteins that are produced would be a formidable blind search. We must have probes—mRNA or protein products associated with such traits—before we can isolate the genes or consider manipulating or transplanting them. So far, no probes for compassion, altruism, musical ability, business acumen, or other such traits are available.

By the end of 1983, in the memory banks of a DEC PDP10 computer in Cambridge, Massachusetts, and in rival computer banks around the world, DNA sequences totaling some two million chemical bases were on line. GenBank—the Genetic Sequence Data Bank in Cambridge—is operated by Bolt, Beranek, and Newman for the National Institute of General Medical Sciences and half a dozen other federal agencies.

Researchers can hook into the database directly by telephoning the computer, or they can buy copies on mag tape updated monthly as new sequences pour in. GGGCGGCGACCT and another 48,502 of the same four letters lined up in a monotonous stream provide the entire genetic recipe for assembling and operating a tiny virus that preys on bacteria—the phage called "lambda." Other monotonous sequences reveal the recipes for an array of normal and abnormal human genes, mouse and rabbit globins, cattle growth hormone, obscure bacterial enzymes, etc.

Isolating critical genetic sequences from all life on earth, dissecting them base by chemical base and cataloging them in computers is the ultimate biological reductionism. This has been called "splitting the genetic atom," and the emerging database is as fundamental to molecular biology as sky maps are to astronomy or the periodic table to chemistry.

While some scientists are cataloging the genetic sequences of living things, others are trying to archive all proteins. A GenBank rival in DNA sequence collection, the National Biomedical Research Foundation in Washington, D.C., has also computerized the structures of more than two thousand proteins since 1965. The DNA banks are now helping to "feed" the protein banks, since protein sequences can be "read" directly from genes. Despite its fifty thousand to one hundred thousand genes, the human genome may code for only a couple of thousand major groups of related proteins. Some biologists believe it won't be too long before all the major types of proteins in the human body are represented in our data banks.

With the development of improved methods, DNA sequencing has mushroomed since the mid-1970s. (In 1977 on-line sequences totaled only about twenty thousand bits.) Sequencing genes is much easier than sequencing proteins, especially proteins that can be isolated only in minute quantities. Once a researcher has a gene, he can clone virtually unlimited quantities of it to work with. Proteins cannot be cloned, and they cannot be synthesized in large quantities in the lab until their amino acid sequence is known. Besides, DNA has only four chemical units to identify;

proteins have twenty. Researchers clone large quantities of a gene, chop it into fragments with restriction enzymes, spread the fragments out on gels that separate them into columns, and read off the sequence directly. A single person can sequence a DNA strand several hundred bases long in a day. The Caltech team is automating this process, too, and gene sequencers should be commercially available soon together with protein sequencers and DNA and protein synthesizers.

Computers are increasingly being used to help scientists make full use of these sequence data. Even with the garble and interruptions removed, the protein-coding region of an average mouse or human gene is a thousand bases long. Finding patterns within a sequence, with or without its garble removed, or comparing several long sequences is next to impossible without the aid of a computer. GenBank and other facilities, as well as researchers themselves, compare new sequences with everything that is already cataloged to look for possible relationships between them.

This comparison is helping researchers to determine the exact chemical differences between activated cancer genes isolated from human tumors and viruses and the almost identical but harmless genes from normal cells. It is also allowing medical scientists to pinpoint the garbles and misspellings in the code that lead to genetic defects so that they can design more effective therapies. Eventually, comparisons of immune system genes and other DNA from various individuals may shed some light on just how different we are from each other biochemically, and what this means in terms of our vulnerability to disease.

No legitimate researcher should have a problem obtaining DNA from many types of diseased or normal cells to sequence and make such comparisons. The Human Genetic Mutant Cell Repository at the Institute for Medical Research in Camden, New Jersey, keeps a frozen library of more than four thousand cell lines stored in liquid nitrogen. The cells, which are available through a catalog, are grown from tissue samples donated by "normal" volunteers as well as by patients with a vast array of genetic defects and chromosome abnormalities.

"DNA cloning and sequencing techniques will eventually give us the whole sequence of the human genome," Walter Bodmer of the Imperial Cancer Research Fund in London told the American Society of Human Genetics in 1980. "Given the present extraordinary rate of technical advance, and the fact that the problem is not conceptual but one of sheer effort, perhaps we shall see the complete sequence within our lifetime.

"But," he added, "whether we shall know what it all means, I very much doubt."

The burgeoning effort to isolate and dissect the molecular anatomy of heredity is after all only half the story. The other half is trying to figure out how it all goes back together and what it all means: where genes sit on the twenty-three pairs of chromosomes; which sits next to which; what sorts of garble and interruptions lie between these protein-coding sequences and what functions, if any, they serve; where the control regions lie; and who decides what genes work in which cells and when.

Tracking genes to specific chromosomes began even before the first genes were isolated. The effort is called gene "mapping." Many of the five hundred genes mapped by mid-1983 still have not been isolated or identified, except by the havoc they cause. By 1970, for instance, genes for eighty human traits such as hemophilia had been assigned to the X sex chromosome by observing their inheritance patterns in families. (Remember, females have two X chromosomes and are usually protected from the effects of any defective genes they carry on one X because the other chromosome carries a normal gene. Males have only one X. Sons who receive an X chromosome from their mothers that carries a defective gene will suffer from it. Daughters who receive the X will be carriers. Diseases passed on in this way are called X-linked.) Only three genes had been assigned to the twenty-two non-sex chromosomes by 1970.

Mapping, like other areas of genetics, progressed rapidly during the 1970s. Researchers have linked some traits to specific chromosomes indirectly by noting that the traits seemed to "travel" from generation to generation in the company of other mapped traits, indicating that the two genes must be sitting close enough together that they seldom get separated in the chromosome shuffle of egg and sperm formation.

Many more traits have been mapped by a process that requires a "forced marriage" between human and mouse cells called cell fusion or cell hybridization. It was one of the first methods developed for transferring genetic material from one cell to another, before the advent of recombinant DNA techniques (see Chapter 13). Such a union sometimes results in a hybrid cell with a full set of mouse DNA but only one human chromosome. If such a hybrid produces a certain human protein, it can safely be assumed that the gene for the protein rides somewhere on that human chromosome. (Groups of hybrid cells that each retain several human chromosomes can also be used for mapping, although the process is more elaborate.) As we noted in gene isolation, however, not all human genes produce their products in a foreign setting unless they are engineered to do so.

More recently, genetic engineering techniques have allowed researchers to map genes directly to specific segments of a chromosome. Using the same hybridization techniques described for gene isolation, scientists can expose separated strands of human chromosomes to radioactively tagged copies of a gene to see which makes a match. Mary E. Harper at the Agouron Institute in San Diego and others have refined the hybridization technique to a sensitivity where genes present in only one or a few copies per cell can be tracked to a specific region of a chromosome.

With techniques like these, the pace of mapping has accelerated so rapidly in recent years that biologist Frank H. Ruddle of Yale University has predicted the whole human genome may be mapped by the end of the century.

Genes for at least fifty inherited human diseases have been mapped to the twenty-two non-sex chromosomes. These include Tay-Sachs and other storage diseases, galactosemia and other defects in carbohydrate metabolism, PKU, sickle cell anemia, the thalassemias, several immune deficiency diseases, Wilms' tumor, and several other hereditary cancers.

Mapping techniques allowed researchers to track the movement of a cancer gene from its normal spot on chromosome 8 to a site on chromosome 14 in patients with Burkitt's lymphoma. The scientists were also able to use these procedures to look at what else inhabited the cancer gene's new neighborhood and find clues to what might have made the gene go "bad."

Investigation of the genetic neighborhood may also help pinpoint exactly which genes are responsible for the mental retardation, heart defects, and other abnormalities suffered by Down's syndrome patients. These individuals are known to carry an extra copy of chromosome 21 in each of their cells, but medical researchers want to know exactly which genes and protein products, in a 50 percent overdose, cause each symptom. A group at Israel's Weizmann Institute of Science is "walking" the twenty-first chromosome in the region where the responsible genes are known to be. Starting with clones of a gene located in that area, the researchers probe a library of chopped-up fragments of the chromosome, hoping to fish out a copy of the gene with some of a neighboring sequence attached. They repeat this process over and over, each time hoping to pull out an overlapping piece that takes them one step farther down the chromosome until eventually they have walked the whole region. The tedium of the process has caused some biologists to rename it "trudging" or "down-trodding."

However, all the mapping techniques mentioned so far, like gene isolation techniques, require that a researcher know something in advance about what he is trying to map: what protein the gene codes for, what its chemical sequence is, or the slim likelihood that it travels with or lives near a gene that is already mapped. But none of this information is known for genes involved in hundreds of inherited diseases like Huntington's chorea or cystic fibrosis. To try to pin down traits like these—and perhaps even track more complex characteristics such as behavior and temperament—biologists are devising a unique mapping system. When they have finished, they hope to have dotted the human chromosome map with so many marker signs that no trait passed on from generation to generation can escape detection.

11. Searching for Landmarks

The next decade should usher in an era of increased understanding of many diseases, from the pediatric birth defects to a wide range of chronic diseases. As this understanding increases, our ability to prevent diseases will follow. Modern medicine may become the art of predicting and preventing diseases in predisposed individuals rather than attempting heroically to prolong life after irreversible damage has begun.
—MARK SKOLNICK, University of Utah, 1980

Southeast of Utah's Great Salt Lake, between the towering Wasatch Range and the barren Oquirrh Mountains, lies an oblong desert valley. Brigham Young and his Mormon followers, driven out of Missouri and Illinois for their religious practices, came upon the arid basin in 1847 and proclaimed it Zion.

The settlement they founded has mushroomed into a modern metropolis of more than half a million people, but the Mormon church still dominates the life and culture of Salt Lake City and other valley communities. Large families are common and strongly encouraged. The church's links with the past are preserved against all earthly destruction in vaults deep beneath the Wasatch. The genealogy records stored there represent the world's most extensive ongoing effort to catalog the human family tree.

For geneticists the valley has become a rich resource. University of Utah researchers, with the cooperation of the church, have computerized the genealogy of seven generations of descendants of the Mormon pioneers, representing 170,000 families. The data bank is being used (with great care to protect individual privacy) for large-scale genetic detective work to spot families with un-usually high rates of cancer, diabetes, heart disease, and other illnesses, rates that indicate a possible genetic vulnerability.

The families benefit directly. Individuals in these high-risk clans can be monitored closely by their physicians and counseled about life-style changes that may keep them from falling prey to genetic predispositions. The geneticists also get an opportunity to look for biochemical signposts like HLA, blood type, or enzyme markers that may signal which family members actually inherited a disease gene.

Only a relative handful of biochemical markers have been iden-tified, however, and few of the thousands of known genetic diseases and susceptibilities will probably be linked to them. The search for a broader range of biochemical markers is continu-ing, but the greatest hope of researchers lies in the genes them-selves.

Using recombinant DNA techniques and with the cooperation of a number of large Utah families, researchers are filling up note-books with genetic fingerprints, and are searching for DNA land-marks that may eventually provide a way to track the passage of virtually any inherited disease or vulnerability through families.

As we saw in Chapter 5, chromosomes do not travel between gen-erations intact. When sex cells are being formed, the twenty-three pairs of chromosomes line up with their partners and prepare to separate so that each egg or sperm will get half a set. (Mating, of course, restores the proper number.) But this is no formal, unem-otional parting. Each threadlike pair may entwine arms and swap remembrances—perhaps one hundred thousand chemical units of one partner's DNA for the equivalent one hundred thousand of the other's DNA. This is called "crossing over," and an average of one crossover per chromosome occurs every time an egg or sperm cell is formed. If a pair swaps two chunks of DNA, it is termed a double crossover.

This shuffling process is the source of each person's genetic and biochemical uniqueness. It means that even if genes for several of Grandpa's or Grandma's traits are linked—that is, carried on the same chromosome—they probably will not be passed down to-gether. A woman, for instance, may inherit one chromosome from her mother that carries genes for blue eyes, auburn hair, and cystic fibrosis. The equivalent chromosome she gets from her father may carry genes for brown eyes, black hair, and Wilson's disease. Because of crossovers, any child the woman bears may receive a chromosome carrying any combination of those grand-parental traits: blue eyes, black hair, Wilson's disease; brown eyes, auburn hair, cystic fibrosis, etc. (Remember, a child must inherit two copies of a recessive gene like the one for cystic fibrosis or Wilson's disease to suffer the illness. We are all thought to be carriers for as many as five to ten potentially lethal recessive genes.)

The closer two genes sit on a chromosome—that is, the tighter the linkage—the less probable they will be separated, and vice versa. Since 1970, as the number of human genes mapped to specific chromosomes by other methods has increased, researchers have been able to assign dozens of additional traits—many of them diseases—to this map by linking them through inheritance patterns to genes already pinpointed.

But this type of mapping tells us only the relative location of a gene on the chromosomes, not whether any specific individual inherited the "good" or "bad" form of it from either parent. Even if we knew on which chromosome the gene for cystic fibrosis is located, and we don't, we would have no way of knowing if a child had inherited Grandma's bad copy and thus, like Grandma, was a carrier. Most recessive traits, like Mendel's wrinkled peas, simply disappear in a family until two carriers marry and both happen to pass along the same disease gene to a child. Even a dominant inherited disease like Huntington's chorea remains silently present and is currently undetectable until the gene begins to express itself by causing neurologic deterioration in midlife.

This isn't true, of course, for all inherited diseases. In about three hundred diseases, visible chromosome defects or abnormal proteins in the blood or other body fluids can betray the inheritance of a disorder before birth or before the onset of illness. In a few disorders, as we will see shortly, scientists can even probe a person's DNA directly for evidence of a defective gene. But for more than three thousand other genetic diseases, neither the gene nor its abnormal protein is known. Instead of waiting for the causes of each genetic disease to be uncovered, some scientists are trying to take a different tack: to find a way to determine in families with a history of genetic disease who inherits the anonymous "bad" genes.

What geneticists want is a tracking system that will let them follow identifiable chunks of Great-Grandma and Great-Grandpa's chromosomes as they break up and shuffle and recombine through the generations. In 1978 biologists David Botstein of MIT and Ronald Davis of Stanford came up with a proposed system that should allow virtually any genetic disease or trait to be tracked through a family, even if nothing is known about the gene involved or the proteins it produces.

The idea is to dot the human chromosome map with genetic landmarks set at regular intervals. Each landmark is to be chosen for the number of different varieties—think of them as colors—it comes in. In any normal individual, a hypothetical marker number 11A on chromosome 11 might be red, or blue, or yellow, or perhaps a half dozen other choices. Signal 11B farther down the line would also come in a variety of forms and so on. The landmarks are actually genetic sequences, but for tracking purposes no one is interested in what their functions are. They just have to be properly spaced, available in a variety of forms, and easy to spot.

With enough of these landmarks in place on all the chromosomes, almost any gene that researchers are interested in following should be sitting close enough to one of them to travel in its company most of the time. The closer a gene is to one of these markers, the less frequently the two will become separated in the passage between generations. The trick to tracking any disease in a family would be to identify which color or form of landmarks it has been riding with: Great-Grandma's red 11A or Great-Grandpa's yellow 11A, and so on.

If a dozen members in four generations of a family developed Huntington's disease, for example, and all of them—but no one else in the family—had a red 13C marker and a green 13D marker, it could be assumed that the Huntington's gene was located near those landmarks. Any child of a Huntington's victim in that family who inherited a chromosome 13 carrying the hypothetical red landmark at the C location and a green one down the line at location D could be almost certain that he inherited the disease gene, too. In another family the mutant gene, contributed to the lineage by a different patriarch or matriarch, might be traveling in the company of different forms of those landmarks.

It is crucial that the markers commonly appear in as many variations as possible. If only red or blue forms are usually found at

13C, for example, and green and yellow at 13D, a number of people in the lineage may be passing along chromosome 13s with the red-C–green-D combination. It would be difficult to tell from the presence of those landmarks alone whether a child had inherited that chromosome chunk from the sick or the healthy side of his lineage. And of course the more DNA markers researchers can eventually place along each chromosome, the more closely they can pinpoint and track the inheritance of any trait.

The idea holds such striking promise that since 1978, scientists around the world have joined the search for landmarks. They are not really landmarks and do not come in colors. Their technical name is "restriction fragment length polymorphisms" or RFLPs. The name is a mouthful, but the concept is simple. Polymorphism comes from the Greek *poly*, meaning many, and *morph*, meaning form. Genetic sequences that are polymorphic exist in a variety of forms, just what is needed for this project. Remember, each restriction enzyme acts like a chemical scissors, chopping the DNA in two wherever it recognizes a single predetermined sequence. Different forms of a polymorphic gene get chopped into different fragment lengths because their sequences are different. Thus RFLPs are simply the various-sized pieces that result when a scientist chops up the landmark DNA with an enzyme. It is size and not color that researchers are using to track these markers through families.

Botstein and Davis were not the first to think of using restriction enzymes to spot the presence of inherited disease. The first application was more direct than the RFLP mapping plan, but also more limited. The idea was simple: If the disease gene is known, and the mutation, the polymorphism, that sets it apart from its normal counterpart has been identified, why not see if one of the 150 known restriction enzymes recognizes that sequence change and cuts the normal and abnormal genes differently. It works.

Sickle cell anemia is one of the best understood of the genetic diseases. A single swapped molecule in the DNA sequence of one of the hemoglobin genes can make the difference between health and a painful, life-threatening illness. The disease itself or the carrier status of children and adults can be detected by taking blood samples and testing the sickling properties of the red cells. In 1974 some researchers began applying these blood tests to prenatal diagnosis, using an instrument called a fetoscope inserted through an incision in the woman's abdomen to withdraw samples of blood from the unborn child. The diagnostic test works, but in about 5 percent of the cases, drawing the blood brings on a miscarriage.

To develop a safer test, researchers needed a way to spot the disease gene directly rather than having to watch for its defective product in a blood cell. This would allow them to probe DNA from cast-off fetal skin and lung cells gathered by amniocentesis—a much less dangerous procedure than drawing fetal blood. Restriction enzymes looked like a promising way to go.

In 1978 Yuet Wai Kan and Andrée Dozy of the University of California in San Francisco used an enzyme called Hpa I to fragment DNA taken from both sickle cell victims and normal individuals. When they probed the resulting fragments for the ones carrying the hemoglobin gene, they found the gene on three different lengths of DNA. The longest fragment carried the sickle gene 87 percent of the time.

Since the long fragment produced by Hpa I did not correlate with the presence of the mutant gene 100 percent of the time, the researchers had to back up the test in some cases by studying fragment patterns produced by the parents' DNA. It was a costly, tedious, and time-consuming process, but it was safe and accurate enough to bring prenatal diagnosis for sickle cell anemia into standard clinical use at a few major medical centers in the United States.

Then in mid-1982 groups led by Kan, Stuart Orkin of Harvard, and John Wilson of the Medical College of Georgia announced a faster, simpler test that could be put to wide use in hospitals around the world. This test uses a restriction enzyme called Mst II to chop the DNA, and the longest fragments produced from these cuts are a sure sign a child has inherited sickle cell anemia.

Even as the work with restriction enzymes was going on, Kan and others were looking for more direct ways of spotting disease genes in an individual's DNA. In 1983 a team led by R. Bruce Wallace of the City of Hope Research Institute in Duarte, California, announced the creation of an artificial genetic sequence that could probe an individual's DNA and latch onto a defective sickle cell gene if it was present. They used the same type of short, synthetic DNA probe—made by hand or by gene machine—that others are using to isolate new genes.

Wallace's group stamped out synthetic copies of short sections of both the normal and sickle hemoglobin gene, tagged them with radioactive signals, and exposed DNA first to one and then to the other. The DNA from normal individuals preferred to combine with the normal probe; that from sickle cell anemia patients preferred the mutant probe. The DNA from sickle cell carriers combined with both probes, indicating the presence of one normal and one sickle gene.

Wallace, Kan, and researchers in Italy have also applied these synthetic probes to detection of a type of thalassemia. Another group has used it to spot alpha-1-antitrypsin deficiency, which leads to emphysema. Still others are working on probes for cancer genes as a potential way of identifying people with an inherited vulnerability to cancer. Researchers are using both probes and restriction fragment length variations to redefine the biochemical HLA markers in more precise terms, at the level of the genes that produced them. In 1982 a group in England reported a possible DNA marker for the still unidentified muscular dystrophy gene.

Precise as these methods are, they still apply only to the few diseases or situations where the defect in or the location of a gene is known. At the time Kan's team was having its first successes with the restriction enzyme Hpa I in sickle cell anemia, Botstein and Davis were brainstorming about a way to use the handy enzymes in detecting a wider range of inherited defects. The idea they came up with was RFLP mapping.

The first step in building the RFLP landmark system is to locate polymorphisms. The next step is to map them, find out how they line up along the chromosomes and what the statistical likelihood is that any two markers will be separated by crossovers. The final step is to try to find links between any of these landmarks and the inheritance of a particular disease.

The polymorphisms Botstein and Davis had in mind are not the rare, often subtle mutations that cause disease. They wanted to tap the rich array of genetic variations that all normal people carry—expressions of the genetic individuality that Garrod at the turn of the century suspected we all possess: different ways of coding the same information, getting the same job done.

But how does one begin looking for things of unknown function that come in a variety of sizes? Botstein and Davis decided to begin by fishing the human gene pool at random. Two colleagues, Raymond White (now at the University of Utah) and Arlene Wyman at MIT, began the task. For probes to fish with, they started with a library of human DNA that had been chopped into fragments and each piece prepared for cloning by Thomas Maniatis, now at Harvard University. For the normal gene pools to fish in, they took skin punches from volunteers around the lab and extracted the DNA.

Extracting DNA is simple kitchen chemistry. A researcher puts cells into a glass tube and pours on detergents to eat away the outer cell walls plus enzymes to devour the proteins. More chemicals and a spin in a centrifuge allow everything but the DNA to be poured away. A dose of ethyl alcohol sends the DNA precipitating out of solution. After chilling the tube in a bucket of crushed ice, a researcher can insert a glass rod, twirl it, and the sticky, mucuslike fibers of DNA will begin to spool onto it like spaghetti.

To spread DNA out for study, a researcher cuts it with a restriction enzyme, places it on a gel sheet, then passes a current through the gel to separate the DNA fragments according to size. The separated DNA is then blotted onto four- by six-inch sheets of filter paper. These so-called blots with their invisible columns of DNA then become the targets for repeated probing.

White and Wyman prepared three dozen random probes from Maniatis's library, then tried to see what size fragments each one would match up with in the blots. This probing is done with the same hybridization technique used for gene isolation. Each blot is put into a kitchen-type Seal 'n' Save bag with quantities of a radioactive probe and incubated for a day or two. Then the excess probe is washed away. A sheet of X-ray film is placed over the paper and the package is left sitting at −70 degrees Celsius for a few days. The result is a clear sheet of film covered with columns of dark bands and smudges—an autoradiograph or fingerprint. The positions of the dark bands reflect the sizes of the fragments the probes pair with and "light up" on the film.

The films with their fingerprints are a permanent record. The blots can then be stripped of their probes and washed in a lasagna dish in preparation for the next hybridization. The DNA remains bound in place on the paper, reusable through dozens of probings.

White and Wyman took their cloned DNA probes and began the tedious task of hybridizing each to numerous blots, looking for ones that would light up a variety of fragment sizes in different people's DNA. They didn't have to wait long. Probe number nine—like all the rest, a DNA sequence of unknown function—lit up at least a dozen variant forms, the longest fragment twice the size of the shortest. The variety, it was found later, stems from the presence of jumping genetic elements, bits of DNA that rearrange to form variant sequences at that site.

That first polymorphism has since been dubbed the "magic marker," and its discovery convinced many skeptics that an RFLP map was feasible. Labs around the world rushed into the search.

"There's no doubt in my mind that this approach will work," Wyman says. "The variation is there in the population. The technology is hard, but it's do-able. And people are interested, so there are more and more people joining the effort."

How many more landmarks are needed for a complete map? White says a hundred or so evenly spaced markers will do for a start. Since the search for the RFLPs is largely random, however, no one knows for certain how many hundreds will have to be found before a properly positioned set of a hundred emerges.

More than a hundred variant sites had been located by 1983, but many are less than optimal. Most of the sequences, for instance, are found in only two forms. Not all the RFLPs have come from sequences picked at random from DNA libraries. Six or eight frequently occurring normal variations have been found at the site of one set of hemoglobin genes, for example. Unlike the hundreds of known mutations at this site that cause diseases like sickle cell anemia and the thalassemias, the sequence changes in or near these hemoglobin genes that result in different restriction enzyme cutting sites seem to be normal quirks that do not cause disease.

With the landmarks beginning to accumulate, White has already started tracking the passage of these known markers between generations to find out how tightly each is linked to neighboring markers on the same chromosome. Physical mapping techniques such as hybridization are developing so rapidly that researchers can use them to localize each of the markers along a specific chromosome and get a rough idea of how far apart they are. But this information isn't enough to predict how often the markers will be inherited together. There is still no firm proof that crossovers between human chromosome pairs are actually random, White notes. Although no hot spots for swapping have been found, studies of the inheritance patterns of markers in large families are needed.

Such studies will also provide something physicians will need when they begin using markers to diagnose genetic disease and provide genetic counseling to families—mathematical odds, data that show landmarks A and B on chromosome 6 become separated only 5 percent of the time, but B and C get shuffled 30 percent of the time. Those are the odds a geneticist will be able to give an individual or the parents of an unborn child at risk of having inherited a genetic disease.

Linkage studies are a classic technique in genetics. In the first half of this century, even before researchers knew what genes were or had any way to study them directly, they used the technique to map the genes of fruit flies. They determined the order and relative distance between the genes for leg length, wing shape, eye color, and other fly traits by painstakingly breeding generations of flies and keeping detailed statistical records of how often various combinations of traits were passed along together. People's life-spans, as well as their insistence on choosing their own mates, make them less convenient to study than flies with their fourteen-day life-cycles.

The large families and stable population of Utah, however, have made it possible for White to find enough volunteers—families with eight to twelve children and four living grandparents—to begin tracking the fate of the RFLP landmarks through three generations. Blood samples are obtained from each family member, and the white blood cells are grown in culture dishes. The DNA from those cells is blotted and probed repeatedly, leaving researchers with notebooks full of genetic fingerprints. The same families will serve as reference families, and each new landmark identified anywhere in the world will be used to probe their DNA.

The researchers can tell something about the company their markers are keeping by looking at each family's films, but the real statistical work is done by computers. White's team started its analysis with five RFLPs on the short end of chromosome 11. ("It just turns out that 11 fortuitously has some of the nicest markers that have so far been discovered," he notes.) Four of them, all worked out by other labs around the country, are from known genetic sites—fragments that carry a type of hemoglobin gene, insulin genes, parathyroid hormone genes, and a cancer gene called *ras*. The fifth site, discovered at the University of Utah, is called ADJ762. Its function is unknown, and for RFLP mapping it is unimportant.

An evenly spaced set of three markers from the short end of

chromosome 11 will be enough to include in a thorough, eco-
nomical-to-use screening network, White believes. But the two
leftovers and any of the extra markers found for other chromosomes will not go to waste. If an initial screening turns up an
apparent linkage between a disease and a particular area of a
chromosome, researchers can go back and screen a second time
selectively, with all the other markers known to be in that region.
This should eventually make it possible to link a disease gene to a
trackable landmark from which it almost never gets separated.

Once the positions and relationships of the markers are determined, the next step is to try to prove guilt by association. Instead
of looking at reference families, researchers will use DNA from
families known to be carrying a disease gene and look for some
link between the appearance of illnesses, such as cancer and heart
disease, and the inheritance of one of the landmarks.

The RFLP network, like the HLA system and other biochemical markers, may start proving its usefulness even before it is all
in place. Data on certain chromosomes, like chromosome 11, may
be complete before any markers have been identified on others,
and many labs are focusing their marker searches in chromosome
regions known to be associated with specific diseases. Teams led
by Uta Francke of Yale and Robert Williamson of the University
of London are zeroing in on the short arm of the X chromosome
where evidence indicates the presence of the gene for Duchenne
muscular dystrophy. Other groups are seeking markers for cystic
fibrosis and Huntington's chorea.

Mapping traits like muscular dystrophy to specific genetic
neighborhoods will help in the eventual isolation of the disease
genes themselves. Researchers can, as we have seen, walk chromosomes, repeatedly probing with known sequences to pull out
overlapping bits of neighboring DNA until they find the gene responsible for the disease.

White also sees another convergence between RFLP mapping
and the isolation of genes thought to cause human disease. Cancer
genes isolated from human tumor cells, for instance, can turn
mouse cells in a lab dish malignant. But do those genes actually
cause cancers in humans? It is unthinkable to transfer cancer
genes into a person to find out for sure. The only way to confirm
the role of these genes may be indirectly, by using nearby markers
(or the cancer gene itself if it is polymorphic like *ras*) to track the
passage of cancer genes through a family with a history of the
disease.

The RFLP network may also give scientists their first opportunity to follow more complex genetically influenced traits. "There are diseases that have something to do with what we call 'constitution,'" Wyman says. "Some people seem more susceptible to some diseases than others. At this point, all we can do is look at statistics and try to make correlations. But with a very good genetic map, we could start to separate out the genetic components from the environmental factors."

This marker system will also enable us to examine for the first time whether mental, behavioral, and emotional traits like compassion, competitiveness, shyness, creativity, and intelligence have a strong enough genetic base to be tracked in families. If Great-Grandpa's tendency to ill humor or his knack for mechanical things are in his genes, the RFLP network should provide a way to spot them.

Mark Skolnick, director of the genealogical database at the University of Utah and a principal in the RFLP mapping effort since the beginning, predicted in 1982 that enough markers will be in place within two to three years to map a large number of diseases. That figure should grow to hundreds of diseases in the next decade or two, he added, and it is time for the health care system to begin gearing up to use this new approach.

"Medicine is becoming a science of prediction and prevention rather than diagnosis and treatment," Skolnick told a congressional committee hearing. "This will have an obvious social impact—longer and healthier lives. It also will require a reorganization of health care whose main forces—insurance companies, hospitals, and doctors—are all tuned to treating late stage diseases in individuals. Much of our national health care budget needs to be reorganized to predicting and preventing diseases in families.

"Is it better to treat hemochromatotics [those who have an inborn iron-storage disorder] for liver cancer, heart failure, and diabetes after they have suffered irreparable organ damage, or set up a national screen for the disease? Insurance companies in general won't pay for screening of genetic diseases, even in high-risk families. Such screening is also not in the interest of hospitals. Doctors often fail to perform follow-up studies on the relatives of individuals with genetic diseases.

"There is a revolution in our ability to predict who will be affected with what disease. It is not too soon to begin considering what institutional changes this revolution implies."

12. Screening the Blueprint

> It is the destiny of modern man to live with enhanced
> power resulting from scientific achievement. He will be
> wise to resist the notion that anything that can be done
> ought to be done. Even so, radical renunciation of power is
> improbable, and it can be as irresponsible as the lust for
> power. Learning to live with unprecedented power is
> probably the most difficult of the tasks that contemporary
> man has set for himself.
> —ROGER L. SHINN, Union Theological Seminary, 1972

Two dozen tiny glass vials filled with clear liquid jiggle in their rack as a young technician slides the assembly into place and flips a switch on an undistinctive machine. In size and shape, the instrument could be a dishwasher. However, the logo on its front panel reads "Gene Analyzer." The vials contain DNA extracted from the cells of fetuses still in their mothers' wombs.

Inside the activated machine a microprocessor starts directing an automated chemistry lab that chops and spreads and readies each batch of DNA for fingerprinting. With tireless efficiency the instrument begins a repetitive ritual: bathing columns of DNA with fluorescent probes, scanning the results, washing away the probes, and beginning again. As probe after probe finds and clings to a matching sequence in the DNA columns, its position is automatically noted and logged. There are probes for hundreds of possible mutations in hemoglobin genes, glitches that lead to sickle cell anemia, the thalassemias, and other blood disorders; mutations in genes for unpronounceable enzymes that cause Tay-Sachs, Lesch-Nyhan, and other devastating syndromes; in genes for structural materials, messenger chemicals, regulatory signals, and so forth.

When the analyzer has hummed to a halt, the genetic anatomies of two dozen unborn children have been probed, cataloged, and printed out. Any inherited diseases are noted, along with susceptibilities to cancer, diabetes, allergies, and other major disorders. The technician removes the printouts and marks them for delivery to the hospital's genetic counseling center where two dozen sets of expectant parents will come to hear the news and talk about their options.

You won't find a gene analyzer in any hospital today, but it is already being designed in the labs of researchers like Leroy Hood of Caltech and Robert S. Ledley of Georgetown University Medical Center in Washington, D.C. No scientific breakthroughs are required. The technology of genetic fingerprinting is already being automated by Hood's lab and other centers for use in DNA sequencing machines. Adapting the process to a clinical lab machine that can be operated by the same technicians who run blood and urine analyzers is an engineering and design problem that will take a few years of tinkering. And direct probing of DNA for defective genes or genetic markers is already being done for a handful of inherited diseases. No one in the field doubts that the number of diseases detectable by these methods will mushroom during the 1980s. The genetic analysis described above is fiction, but only in its speed and scope. The technology is in hand.

What these developments portend is dramatic growth for a type of preventive medicine that was almost unheard of two decades ago: genetic screening and counseling. It promises to influence not only the children we bear, but how and where we live, who we marry, and what jobs we hold. Expanded genetic screening will be the first major impact of recombinant DNA technology on health care, and we will be able to detect most defects long before we are able to treat them at the genetic level.

Since 1967, when amniocentesis was used for the first time to diagnose a chromosome abnormality in a fetus still in the womb, prenatal diagnosis has matured to become a routine option in obstetrical care for high-risk mothers—so routine that damage suits have been won by parents of some retarded and genetically afflicted children against physicians who failed to inform them that prenatal tests were available. With these tests and the option of abortion, tens of thousands of at-risk couples have been able to prevent the birth of children with genetic defects and, more important, bear unafflicted children of their own.

By 1981 some two hundred genetic diseases and fetal abnormalities and virtually all chromosome defects such as Down's syndrome were diagnosable before birth, although many of the tests for specific diseases were available in only one or two major medical centers. Advances in genetics promise not only to expand this list, but to make diagnostic techniques faster, safer, cheaper, and more widely available.

This partial escape from the genetic lottery means parents must accept responsibility for an ever-widening range of decisions about the kinds of children they want to bear. With diseases such as Tay-Sachs that destroy the brain and kill a child in the first few years of life, most families find a rather clear-cut moral choice. In the future, on the other hand, as the list of detectable vulnerabilities extends to allergies, arthritis, diabetes, hypertension, schizophrenia, cancer, and other disorders, pregnancy may become a time for couples to examine their most deeply held hopes and values. We may all, as families and as a society, have to make decisions about just how much biochemical variation and physical imperfection will be "normal" and acceptable in future human beings.

Fetuses are not the only subjects for genetic screening. Since the early 1960s, when the first inexpensive mass screening method for detecting PKU in newborns was developed, infant screening tests for two dozen other disorders have been put into use. Virtually all states have newborn screening programs for one or more of these disorders—PKU, hypothyroidism, homocystinuria, maple syrup urine disease, galactosemia, sickle cell anemia, and others. In PKU and some other diseases, accurate early diagnosis and treatment can prevent mental retardation or other suffering. Diagnosis of conditions such as cystic fibrosis affords no cure yet, but does give families the option of avoiding giving birth to other children who might be affected.

There is a third screening level, still largely visionary and perhaps the most controversial of all: so-called genetic typing of children or adults for inherited vulnerabilities. The tests cannot tell us with any certainty, years in advance, what we will suffer or die from. But they may allow us to avoid specific environmental factors—foods, medicines, liquor, cigarettes, germs, sunshine, pollutants, industrial chemicals, stresses—that our individual genetic constitutions cannot handle well. Genetic typing could also supply us with a list of the hidden recessive disease genes we all carry, something we might wish to consult when considering marriage or children.

For some people the prospect of this type of prophecy represents a new degree of personal freedom and control; for others, a new burden of personal decisions. But most troubling is the issue of who will be allowed access to our personal forecasts: Potential employers? Insurance companies? Public agencies? The possibility exists that we will escape the domination of our genes only to find ourselves under a different set of controls—formal or informal limits on where we can live and work, who will insure or marry us, and so on.

The nation got a taste of how damaging this sort of genetic "Scarlet Letter" can be in the early 1970s when, in an ill-planned effort to do good, Congress and a number of states initiated mass screening of black children and adults for sickle cell anemia. Some states made the programs mandatory and required blacks to be screened before they could receive marriage licenses or attend public schools. Many of the programs failed to guarantee confidentiality of the results or educate the public and the people screened about the difference between actually having the disease and being a carrier of the sickle trait. Prenatal tests for sickle cell anemia were new and not widely available then, so couples who found out they were both carriers had only two options: Cross their fingers during pregnancy or not have children at all.

The result was great confusion and fear in some black communities, about the disease itself and the motivations of the whites who were so eager to detect it and offer advice on reproduction to blacks. There were some reports of carriers being denied jobs or insurance.

Advances in molecular biology, by expanding the number and types of genetic tests available and undoubtedly sparking greater demand for genetic screening at all levels, will force us to face such personal, social, and legal issues on a broader scale than before. Future mass screening programs will have to be carefully designed to ensure that they serve the needs and hopes of the people being screened. As we will see later, one highly successful model for future efforts is already in place—a program for the detection of Tay-Sachs carriers.

In the 1950s only a handful of genetic counseling centers existed in the United States. All that counselors could offer was mathematical odds. If a couple had already borne a child with cystic fibrosis, sickle cell anemia, Tay-Sachs disease, or another recessive disorder, it was obvious they were both carriers. A counselor could explain Mendel's laws to them: With each pregnancy they faced a one-in-four chance of producing another afflicted child. They also had a one-in-four chance of producing a completely normal child, and two chances in four of producing one who, like them, was a carrier. If one parent had a dominant disease such as Huntington's, each child had a fifty-fifty chance of having it, too. Couples who still wanted to bear their own children had to play roulette.

The development of amniocentesis took medical genetics beyond gamblers' odds for the first time and gave physicians a window to the unborn. A German researcher actually devised the technique in 1882 to relieve an excess of amniotic fluid during pregnancy. It was largely forgotten, however, until the 1950s when medical researchers resurrected it, using it late in pregnancy to survey fetal damage caused by certain nongenetic diseases. The possibility of using the procedure for the diagnosis of inherited disease opened up during the 1960s because of developments in another field: Cell culture techniques were advancing rapidly, and researchers found they could coax the few fetal cells found in an early amniotic fluid sample to grow and multiply into quantities large enough to be tested.

In 1967 and 1968 a few researchers reported success using amniocentesis to diagnose Down's syndrome and other chromosome abnormalities as well as inherited diseases like galactosemia. The impact was immediate. Newspaper and television coverage of these successes was widespread, and often sensational. It came at a time when cultural attitudes and state laws on abortion were being liberalized and attitudes toward family planning, family size, and personal choice in medical matters were changing. The requests from couples who wanted their pregnancies monitored began to increase steadily.

Amniocentesis itself is not a diagnostic test. It consists of drawing a sample of amniotic fluid and cast-off fetal cells from the womb for testing between the fifteenth and seventeenth week of pregnancy. The fluid itself is checked for elevated levels of alpha fetoprotein (AFP), which indicates the possibility of major defects in the fetal brain and spinal cord such as spina bifida. (Test kits for measuring AFP levels in maternal blood are now available to physicians, but their use is controversial since positive results are not reliable and must be rechecked by amniocentesis.) The fluid or the cultured fetal cells can be tested for the enzyme deficiencies found in Tay-Sachs and galactosemia. Diseases like familial hypercholesterolemia and Lesch-Nyhan syndrome can be spotted by watching how the cells bind or take up certain radioactively labeled chemicals.

Chromosome abnormalities, such as Down's syndrome, are checked by lining up, counting, and scrutinizing the chromosomes microscopically—a procedure called karyotyping. Machines for fully automated chromosome analysis are even further along in development than gene analyzers. The fetal chromosomes can also reveal the sex of the fetus. For many X-linked diseases like Duchenne muscular dystrophy in which there are no reliable prenatal detection tests, the only way families can be sure of having an unafflicted child is to bear only daughters. Half the daughters will be carriers, but half the sons will be sufferers. (Amniocentesis for sex selection where no genetic disease is involved is universally discouraged. Physicians report few cases of couples wanting to undergo the test and consider a midtrimester abortion simply to ensure a child of a preferred sex.)

The DNA from fetal cells obtained at amniocentesis can also be probed directly for the presence of the mutant genes that are involved in sickle cell anemia, some thalassemias, and a deficiency of the enzyme alpha-1-antitrypsin that leads to pulmonary emphysema. As more disease genes are identified, or genetic markers for disease genes located, this type of prenatal testing will become more common.

Amniocentesis is only the first of several procedures that are revolutionizing prenatal diagnosis, and researchers are already at work on new techniques that will eventually replace amniocentesis. Ultrasound—bouncing high-frequency sound waves off the fetus to generate an image on a TV screen—is now routinely used along with amniocentesis to help locate a safe spot to insert the needle, determine fetal age, look for the possibility of twins, and check for major structural defects in the fetus.

Another diagnostic procedure, fetoscopy, allows researchers to insert a slender periscope through a tiny incision in a pregnant woman's abdomen and look directly at a small portion of the fetus—such as fingers, eyes, or genitals—for abnormalities. Fetoscopy also lets physicians take samples of fetal blood or tissue for study. Until methods of testing DNA directly were developed, fetal blood samples provided the only way to diagnose hemoglobin diseases like sickle cell anemia and the thalassemias before birth. This is because the cells retrieved in amniotic fluid are mainly lung and skin cells, and proteins produced exclusively by the blood, brain, liver, and other specialized cells cannot be detected in amniotic cell cultures. Scientists are looking for ways to make amniotic cells more useful, perhaps by fusing them with specialized cells and forcing them to turn on dormant genes such as hemoglobin so that their protein products can be inspected for defects.

Fetoscopy is still largely a research technique, used at only a few major medical centers in the United States. When it was first used in human pregnancies in 1973, fetoscopy led to miscarriage 5 to 10 percent of the time. Although now the rate of miscarriage in experienced hospitals is closer to 3 to 5 percent, the risk to the unborn from this invasive technique is still high. In contrast, several large-scale studies of amniocentesis have found no significant risks to the mother or fetus when the tests are done in experienced medical centers.

In the future, however, the ideal sampling technique for prenatal diagnosis will require no invasion of the womb with needle or scope—perhaps it will involve only a blood sample from the mother. And it will be done months earlier than amniocentesis can be performed. By the second month of pregnancy, tiny amounts of fetal blood begin to seep into the mother's system. Since fetal blood cells carry their own unique identity markers, some scientists are trying to find an efficient and economical way to separate them out from the mother's blood and culture them for diagnosis.

Another new approach is a procedure called "chorionic villus sampling" that is already being tested on pregnant women in the United States and Europe. Chorionic villi are hairlike projections of the membrane that surrounds a fetus during the first few months of pregnancy. A physician can insert a small plastic tube into a pregnant woman's cervix and, guided by ultrasound, remove a sample of tissue from one of the villi. The sample contains enough fetal cells for immediate genetic analysis, eliminating the several-week delay now required to allow the meager quantity of cells obtained by amniocentesis to grow and multiply in culture.

* * *

While second-generation techniques such as these will probably someday replace it, amniocentesis remains the foundation for prenatal screening today. In the first decade of its existence, 1968 to 1978, the use of amniocentesis increased dramatically. About forty thousand pregnant women in the United States were tested during that period, fifteen thousand of them in 1978 alone. The number of centers offering the service rose from only 10 in 1970 to more than 125 by 1979.

At the end of that first decade, the National Institute of Child Health and Human Development organized a series of task forces and a consensus development conference to assess the status of prenatal diagnosis. In a report issued in 1979 it was concluded that amniocentesis "clearly has moved beyond the level of a research procedure to an accepted part of clinical practice" for the assessment of certain at-risk pregnancies.

Who is at risk? The 1979 report listed a number of situations in which a physician should inform expectant couples that the tests are available and counsel them about their risks of bearing a child with an inherited defect. They should be informed when:

—The woman is thirty-five or older. (Chromosomal abnormalities are much more frequent in the children of older mothers. The odds of giving birth to a child with Down's syndrome, for instance, are 1 in 885 at age thirty, 1 in 365 at age thirty-five, and 1 in 32 at age forty-five. The sperm of older fathers, on the other hand, is subject to higher rates of mutations in single genes, which can lead to the spontaneous appearance of new dominant diseases in a family. But these disorders cannot be detected by chromosome studies or other routine prenatal tests.)

—The woman has previously borne a child with a chromosome abnormality, for example, Down's syndrome or fragile X.

—The mother or father or one of their family members is known to have a chromosome abnormality.

—The woman, or a former wife of the husband, has had three or more spontaneous abortions. (Estimates indicate 50 to 60 percent of miscarriages in the first three months of pregnancy involve fetuses with chromosome abnormalities. Only a small minority of chromosomally defective embryos that are conceived survive to birth.)

—The woman has previously borne a child with multiple serious malformations.

—The woman has male relatives with Duchenne muscular dystrophy, hemophilia, or other X-linked disorders, and thus may be a carrier.

—The couple is known to be at risk for bearing a child with a genetic disorder, such as sickle cell anemia or Tay-Sachs disease, that can be detected prenatally. This risk is known if the parents have already borne an afflicted child or if they have been tested and found to be carriers of the disease genes.

—The fetus is at risk for a brain or spinal-cord defect like spina bifida. There is a risk if the woman has already borne a child with such a defect, if one of the parents has such a defect, or if the mother's blood AFP level is elevated.

It is important to remember that amniocentesis is not recommended for all pregnant women at this point, since it is not yet a mass screening technique. And no woman who undergoes the procedure will be tested for any single-gene diseases like Tay-Sachs or for familial hypercholesterolemia unless she is already known to be at risk. Checking for individual diseases is expensive, and fluid or cell samples for each biochemical test must be sent to one of the few labs around the country that perform each of them essentially by hand. The extremely low chances of spotting an afflicted child, coupled with the cost and the lack of lab capacity, make it impractical to mass-screen every fetus for everything that can be detected. Mass screening also could not guarantee anyone a normal child, since the majority of inherited diseases still cannot be pinpointed before birth.

Eighty to 90 percent of the pregnancies monitored by amniocentesis so far have been screened only for chromosomal abnormalities, usually because of the age of the mother. (AFP tests for brain and spinal-cord defects are also performed now whenever amniotic fluid is drawn.) In 95 percent of the cases, the results are negative and couples can enjoy the rest of the pregnancy without anxiety. When a defect is found, the choice of aborting or continuing the pregnancy is up to the family. As of 1979, abortion was chosen in 95 percent of these cases. In a few instances, as we saw in Chapter 9, prenatal diagnosis has led to successful treatment of the unborn, through fetal surgery, or through special diets, vitamin supplements, or drugs given to the mother. Other diseases such as galactosemia, diagnosed in the womb, can be successfully treated at birth. Few genetic diseases fall in this category presently, but the hope is that in the future many conditions detectable by screening will eventually become treatable.

Making prenatal diagnosis available to every pregnant woman is not the major concern of geneticists and agencies involved in maternal and child health today. Their first goal is to make amniocentesis available to all those who could benefit most from it. Even among women thirty-five or older, who make up the majority of clients for prenatal screening, only a fraction of the pregnancies are being screened. The 1979 report of the National Institute of Child Health and Human Development found that only 5 to 6 percent of the 140,000 to 150,000 births to older mothers each year were being monitored. The number of births to women thirty-five and older was projected to rise to 180,000 by 1984, and social trends toward later pregnancy are expected to accelerate this increase.

Despite the media attention given to the "right-to-life" movement, antiabortion sentiments apparently do not play a strong role in limiting the use of amniocentesis. The 1979 report noted that surveys have consistently found that a wide majority of Americans support abortion in cases when a child may be born seriously disabled. The major factors preventing full use of amniocentesis by at-risk women are the cost; a lack of awareness on the part of the public and even physicians; the tendency for indigent or poorly educated women to delay seeking medical care until the fifth month of pregnancy, too late for the option of abortion; the limited availability of lab and counseling services; and a limited awareness of preventive medicine.

Only since 1980 have some major health insurance companies begun to cover the costs of amniocentesis, and most will still not pay for other genetic diagnostic services or genetic counseling. The technology for probing the unborn, for diagnosing inherited defects, and even for treating them is racing far ahead of its widespread application to those who could benefit from it.

"It will undoubtedly take at least a generation of planned and consistent educational efforts to bring about general concern in society and an intellectual commitment to minimize the burden of genetic disease," the report concluded. "But a concerted effort must begin now if we are to prepare the parents of future generations to deal intelligently and effectively with the problems and potential of new knowledge in human genetics."

The need for education and awareness go beyond fetal screening. One big step in minimizing the burden of inherited disease in the future will be the expansion of carrier screening programs for recessive and X-linked disorders. If at-risk couples who carry the same recessive disease genes are identified, counseling and prenatal diagnosis can be made available before they have given birth to their first defective child.

Tay-Sachs is a cruel disease. A child who inherits it may seem perfectly normal for several months after birth. But fatty lipids have already begun building up in the brain cells, destroying their function. Within the first year, the child stops progressing, then begins to lose ground. He or she becomes weak, listless, and eventually blind and paralyzed as the nervous system deteriorates relentlessly. Most victims die within two to four years of birth. There is no treatment for this disorder.

About 85 percent of Tay-Sachs victims are children of Ashkenazim, Jews of eastern European descent. The majority of this country's Jewish population, as many as six million people, have this lineage. An estimated one in thirty carry the recessive gene for Tay-Sachs. A high incidence of the disease has also been found in some other well-defined groups, including an isolated population in Nova Scotia. Over the past fifteen years, voluntary mass screening for Tay-Sachs carriers has become the first widely successful effort of its kind, a model for future carrier detection programs.

Unlike many of the early attempts at mass screening for sickle cell anemia among blacks, the Tay-Sachs programs have focused on education of the target population, made an effort to involve community leaders, and provided for extensive follow-up and counseling of identified carriers. Also of critical importance has been the fact that reliable tests are available to diagnose the disease prenatally. This means screening programs have more to offer carrier couples once they are identified than gamblers' odds.

During the decade of the 1970s, about 400,000 people in fifteen countries around the world have voluntarily taken a simple blood test for the enzyme involved in Tay-Sachs. Carriers have less than the normal amount of enzyme because they have only one good gene for it. Children with the disease have a complete enzyme deficiency. About 5,000 carriers, including 350 couples in which both husband and wife are carriers, have been identified. During this time prenatal monitoring of more than 1,000 pregnancies of carrier couples resulted in detection and abortion of 250 afflicted fetuses and, more important, in the birth of 800 healthy children. (Studies have repeatedly shown that most families who have already borne a child with a serious defect avoid pregnancy or abort those that occur unless prenatal tests are available.)

Medical geneticist Michael Kaback of the University of California at Los Angeles—who launched a pilot screening program for Tay-Sachs in the Baltimore-Washington area in 1971 and later became head of an international center that coordinates the worldwide detection effort—has reported that the incidence of Tay-Sachs disease in the American Jewish community dropped 65 to 85 percent during the 1970s.

The only other widespread carrier screening effort to achieve any success to date is for beta-thalassemia. In Greece and Italy, where the prevalence of this blood disease is a major economic and health problem, carrier screening coupled with prenatal diagnosis by fetoscopy and fetal blood sampling have been made widely available.

In the United States, 750 children a year are born with sickle cell anemia. This disease is six times more prevalent among blacks than Tay-Sachs is among Jews. Despite the earlier unfortunate attempts to screen for sickle cell anemia in the black population, as new techniques for prenatal diagnosis by direct DNA probing become more widely used, carefully managed voluntary carrier screening is likely to expand.

Other detection programs will undoubtedly be launched as we learn to identify the presence of cystic fibrosis and other disease genes in both carriers and afflicted fetuses. The aim of all these efforts is to supply couples with information to help them make personal family planning decisions. Traditionally, our society has left the choice of what to do with such information up to the families involved. Abortion has been an option for more than a decade to those who choose not to bear children with genetic defects. On the other hand, no move has been made to deny public medical and financial support to afflicted and handicapped children suffering from conditions that could have been detected before birth. In the foreseeable future, society is not likely to try to step in and mandate personal decisions about childbirth—especially when only a small percentage have any options yet about the genetic health of the children they bear.

Not everyone believes, however, that an individual's genetic makeup should be strictly a personal concern. The issue may confront us first in the workplace, not in the obstetrician's office.

*   *   *

The congressional Office of Technology Assessment (OTA) created a stir in the summer of 1982. At a House subcommittee hearing, the OTA released results of a mail survey it conducted of the
Fortune 500 corporations, fifty major utilities, and eleven labor
unions. A key question was, would these firms consider using
some type of genetic testing of employees in the next five years?
Fifty-nine of the major firms said yes. Seventeen said they had
conducted such tests in the past, although only six were currently
using them. The responses were anonymous.

Subcommittee Chairman Representative Albert Gore, Jr., D-
Tenn., expressed alarm. The information he had gathered from
scientists indicated that genetic screening of adults for vulnerabilities to chemicals and other workplace hazards was still a
"black art," he said.

"It is currently considered irrational and wrong to exclude workers as a result of black skin or because they are male or female, but if
one has a genetic makeup making one ten percent more susceptible
to disease, does that justify excluding?" Gore asked. "At what point
is it invidious discrimination to deny a person a job on the basis of
genetic heritage?" Labor leaders joined in expressing concern about
the potential for creating a class of genetic lepers, perhaps weeding
out and refusing to hire the most susceptible people rather than making the work environment safer for everyone.

The OTA survey actually lumped together two types of screening—testing workers for genetic susceptibility to chemical exposures, and checking them for genetic damage caused by such
exposures. Which type of screening the corporations were considering pursuing was not clear, but most scientists agree neither
test is likely to be widely used in industry anytime soon simply
because the findings are not that meaningful.

As we saw in Chapter 9, for instance, the hereditary deficiency
of an enzyme called G6PD can leave individuals vulnerable to
hemolytic anemia when they are exposed to certain drugs and
chemicals such as naphthalene, which is used in making explo-
sives, lubricants, solvents, and dyes. There is no way to predict,
however, which G6PD-deficient people are sensitive to which
drugs and chemicals, and to what degree. Individuals with severe
alpha-1-antitrypsin deficiency will almost certainly end up with
emphysema, and will suffer sooner if they smoke or otherwise
irritate their lungs with fumes and pollutants. Some researchers
suspect that carriers who inherit only one defective gene are also

predisposed to respiratory problems, but there is no firm proof for this. One of every twenty-five to fifty Americans, mainly those of northern European ancestry, may be carriers.

The HLA markers also give us only percentages so far. They cannot pinpoint who will suffer from any particular environmental exposure. One of the tightest linkages found to date is the relationship between the HLA marker B27 and ankylosing spondylitis, a chronic spinal arthritis. Almost 95 percent of those who suffer from the disease carry the marker, but so do 4 to 8 percent of the general population. Most of those with the marker will never get the disease. For those who do, the effects of the disease can range from mild to crippling.

Should a pro football team be allowed to reject people with B27 markers for fear of ending up footing the bill for possible back problems? Should a chemical company have the right to refuse to hire some or all people with G6PD deficiency or an alpha-1-antitrypsin carrier? What about people prone to cancer or heart attacks who may eventually draw on the company's health plan? At present it seems unlikely such attempts will be made. Too few at-risk traits are known, and the odds on them are too imprecise.

However, this won't always be the case. HLA types are now being subdivided and linked with much greater precision to diseases. Others disease susceptibilities will be pinned down using the DNA landmark maps. Genetic typing for a variety of traits eventually will not only be possible but common. Our children and grandchildren will come to expect and want knowledge of their inborn quirks as a way of minimizing their risks and gaining more control over their fate.

Whether anyone else should have access to those data, and what limits others should be able to place on us because of our genetic makeup, are questions society must consider. Insurance companies and doctors performing pre-employment physicals are already privy to our blood pressure, back problems, and personal and family history of epilepsy, heart disease, mental illness, etc. A listing of genetic vulnerabilities may eventually be considered part of any thorough medical record. The issue of disclosure also comes up when relatives of an individual with genetic disease or children given up for adoption could benefit from being informed of their inherited risks. Representative Gore's subcommittee, labor organizations, specialists in occupational health, and several other concerned groups are already actively considering how legitimate needs for disclosure can be handled without violating individual rights to privacy.

Mass screening of fetuses, newborns, and adults will eventually help us cut our health care burden by reducing the numbers of infants born with inherited defects and by allowing us to avoid environmental factors that may trigger disease. But there are still thousands of disorders that cannot be detected until they strike, and new mutations arise with every generation. Despite expanding preventive measures, we are going to need treatments at the genetic level—therapies that replace or correct the workings of defective genes or counteract their damaging impacts.

IV
TRANSPLANTING GENES

13. The Genetic Shuffle

Once we thought the DNA of complex organisms was
inscrutable. Now we cope with it readily.
—MAXINE SINGER, National Cancer Institute, 1980

The stuff of heredity has proved to be surprisingly easy to move
from one creature to another, probably because humans were not
the first to think of it. Apparently nature has been swapping genes
since the early stages of evolution, introducing possibly useful
variety into living things by offering them occasional bits of new
DNA as well as making random glitches (mutations) in their own
instructions.

No one seems to have benefited more from this movable genetic feast than the microbes, an enormously successful class of
creatures that is continually trying out and trading around new
bits of DNA that might provide an edge in the battle for survival—
now largely a battle against man and his medicines. Viruses may
accidentally capture bits of DNA from a cell as they leave and
carry it along into the next cell they invade. Bacteria may take in
stray remnants of genes they run across. Or they may pass among
themselves circular molecules called plasmids that carry one or
more genes, often for resistance to antibiotics, deadly substances
produced first by their natural enemies the fungi and now by man.
(In the past ten years, we have learned to make good use of these
bacterial plasmids as vehicles to ferry genes into bacteria for
cloning.)

It was by observing one of these natural genetic transfers in a
microbe several decades before the advent of modern genetic engineering that scientists first identified the physical substance of
heredity. The units of heredity had been named genes long before
anyone knew what they were made of or how to pinpoint them
amid the chemical ratrace inside the cell.

As soon as scientists figured out what genes were, they began
trying to move them about. It was an inevitable goal. The secrets

of human development—how a creature with a hundred trillion specialized cells blossoms from a single fertilized egg cell—and the understanding and treatment of hereditary disease require a knowledge of how genes work and are regulated in the cells of higher animals. But a single gene is too small to be seen even with our most powerful microscopes, and it is impossible to track the fate and function of one gene among fifty thousand to one hundred thousand others in its natural setting. Researchers needed to find a way to get genes to work in places where they could be watched: alien settings where they felt at home enough to be active but could never melt into the crowd.

The work has moved with amazing speed. Within the past five years, researchers have developed a variety of techniques for putting any gene they can get their hands on into foreign cells or even live animals. Most genes can even be forced or tricked into working in some fashion, although not usually in their natural way.

Most of the individual gene transplant experiments described later in this section were not meant to be a direct path to human therapy. They were designed to answer basic questions about how living systems function. The systems molecular biologists work with—when they move fruit fly genes into frog eggs, bacterial genes into human cells, or rabbit genes into living mice—may appear to have little to do with helping prevent mental retardation or determining your vulnerability to cancer. But no scientist tinkering with mouse cells and obscure viral genes today is unaware that his findings and methods may shortly be applied to man. Many lab-based researchers, who naturally focus on how much is still unknown rather than how much has been learned already, consider talk of gene therapy premature. Twice, however, clinical researchers, encouraged by how much has been learned from gene transfers in animals and impatient at having no other hope to offer suffering patients, have tried to put new genetic material into humans. Neither attempt brought relief to the patients. But every advance in an animal now raises the question among physicians: Is it time to try again?

The scientists who began in the 1960s to try to move hereditary material around had neither specific isolated genes nor recombinant DNA techniques to work with. So they concentrated on trying to move the total DNA content of one cell into another. This turned out to be amazingly easy, although the procedures seem crude and haphazard compared with the gene transfer techniques available less than twenty years later.

One of the first technical tricks scientists devised in the 1960s for getting foreign genetic material into animal cells can be described as a forced marriage: Two different lines of body cells— say, skin cells from a mouse and a man—are mingled in a lab dish and doused with a chemical that seems to overcome their tendency to remain private and aloof. A solution of certain inactivated viruses or polyethylene glycol, a relative of antifreeze, does the job. Under the influence of such agents, some of the cells touch, dissolve the membranes that separate them, merge their contents, and fuse their nuclei. The hybrid cell will contain two full sets of chromosomes, with all the information necessary to make a mouse and a man. The technique is called cell fusion.

Some researchers have even doused embryos from different strains of mice with enzymes to remove their gelatinous protective coatings and forced them to mingle in a lab dish. Those that clumped together and fused formed giant embryos which were implanted into foster mothers to develop. The result was not monsters but normal mice, mosaics made up of populations of cells from two different strains. When black and white strains were fused, the resulting mouse bore a patchwork coat, the color of each patch determined by the origin of that particular clump of cells.

Speculation is that man is not the first to make use of cell fusion. Within the cells of man and other eukaryotes (organisms with nuclei in their cells; bacteria and other prokaryotes have none) exist structures that look suspiciously as if they may once have been separate creatures: the mitochondria, the powerhouses in animal cells where respiration and energy production occur, and the chloroplasts in green plants where the energy of sunlight is used to produce the nutrients on which the earth's food chain is grounded. These organelles may be the descendants of ancient bacteria that formed successful partnerships with primitive nucleated cells. In the primordial seas the bacteria may have been engulfed by nucleated cells. Instead of digesting them, a few of the cells may have let the bacteria set up housekeeping and do chores in return for lodging. Now lodger and host are one, inseparably bound and dependent.

At another point in evolution, when single-celled organisms began assembling themselves into more complex, multicelled creatures, fusion might have provided a way for them to acquire potentially useful new genetic capabilities.

But fusion is imprecise and the resulting hybrids are usually clumsy and unstable. As they divide and produce daughter cells, the hybrids tend to lose some of the parental chromosomes at random. The technique has nevertheless proved valuable, as we have seen in Chapter 10, for mapping genes to specific chromosomes. If, for example, a colony of mouse-human hybrid cells produced by fusion retains only a single human chromosome, and the cells are found to produce a certain human enzyme, then the gene carrying the manufacturing specifications for that enzyme must be located on that chromosome. Hundreds of human genes have been mapped this way.

Another spin-off of cell fusion research is the production of highly specific "monoclonal" antibodies, one of the first commercial and medical payoffs of genetic engineering.

Antibodies are the body's first line of defense. Each is programmed to seek out a specific foreign invader and mark it for destruction. But the body typically mounts a shotgun defense against any invasion, and traditionally researchers have had to be content with harvesting a witches' brew of antibodies from the blood serum of animals challenged by germs or foreign material.

A single antibody-producing spleen cell, grown into a colony in a lab dish, will yield quantities of one specific antibody. However, the cells do not live long in a dish. In 1975 British scientists Georges Kohler and Cesar Milstein of the Medical Research Council in Cambridge, England, came up with a solution to this problem: They fused the spleen cells with a type of mouse tumor cell that produces no antibody of its own. Tumor cells are virtually immortal in culture. The resulting hybrid cells—hybridomas—retain the desirable qualities of both parent cells. They are immortal factories producing a continuous supply of one pure antibody. The antibodies are called monoclonal because all the cells in the colony are clones, identical to the original hybrid.

Researchers are developing monoclonal antibodies that can recognize fetal cells in a sample of a pregnant woman's blood or allow precise identification of HLA types. Commercially, monoclonal antibodies are fast replacing traditional reagents in diagnostic test kits for everything from cancer and infectious disease detection to allergies and pregnancy. The antibodies alone—or linked to toxins, drugs, or radioactive isotopes—are also being tested in the treatment of cancer.

Plant genetic engineers are using cell fusion to try to create hybrids of plant species that could not normally be crossbred—blight-resistant tomatoes with potatoes, fine wine grapes with hardy wild varieties, nitrogen-fixing bean plants with grain crops. The technique is called protoplast fusion because the tough outer walls of the plant cells must be chemically stripped off before fusion can be attempted. The stripped cells are termed "protoplasts." The goal is to produce hybrid cells containing desirable traits from each parent and to use plant-tissue culture techniques to grow whole new plants from these cells. (Equivalent culture techniques do not exist for regrowing animals from bits of body tissue the way geraniums can be regrown from cuttings. Otherwise, livestock breeders might also be interested.) But protoplast fusion, like all fusion, is crude and random. Plant scientists are trying to perfect more precise methods for isolating and transplanting the specific plant genes involved in nitrogen fixation, drought and disease resistance, and so on.

Like plant scientists, molecular biologists have also turned to newer techniques that will allow them to transfer precise and specific bits of genetic information instead of random numbers of whole chromosomes into cells and organisms. In recent years some variations on cell fusion have been developed using specific genes. One approach is to trap genes within man-made fat molecules called "liposomes," or "lipid vesicles," and fuse these with target cells.

Another technique is the microbial version of protoplast fusion—"spheroplast fusion." After a gene is inserted into a bacteria for cloning, it is often amplified, reproduced independently within the bacterial cell. The end result is a bacteria containing fifty to a hundred identical copies of the gene. Instead of removing these genes, purifying them, and then finding a way to transplant them into an animal cell, researchers can simply dissolve the bacterial outer wall with enzymes and fuse the whole package to a mouse or human cell. The drawback, however, is that the resulting hybrid cell receives not only the genes the scientist wanted to transfer, but all the bacterial DNA as well. Neither spheroplast fusion nor liposomes have been widely used yet, and both methods have been overshadowed by the successes of other techniques.

During the 1970s three major methods were developed for delivering isolated genes into animal cells in a lab dish or into living animals. Two were variations on natural gene-swapping processes observed in microbes over the past fifty years. The first method makes use of infectious viruses to ferry foreign genes into animal cells. A second allows scientists to transform cells just by incubating them with naked DNA. The third technique is a human innovation—using microscopic needles to inject genes directly into individual cells or even into embryos.

14. The Friendly Virus

We live in a dancing matrix of viruses; they dart, rather
like bees, from organism to organism, from plant to insect
to mammal to me and back again, and into the sea, tugging
along pieces of this genome, strings of genes from that,
transplanting grafts of DNA, passing around heredity as
though at a great party.

—LEWIS THOMAS, 1974

Viruses exist in the biological twilight zone between life and non-life. They are not cells. Without us—and by "us" I mean all the cellular creatures on earth, from bacteria to humans—they could not live. A virus is simply a packet of information, a string of genetic material (either DNA or RNA), a template sealed in a protective capsule. Unlike living cells, which contain complex machinery for copying their genetic blueprints, processing nutrients, using energy, manufacturing proteins, and reproducing themselves, viruses can do nothing on their own. They are the ultimate parasites, completely inert, lifeless without a host. For a virus, infection is life.

To multiply and perpetuate itself, a virus cannot grow and divide like a living cell. It must invade a cell, strip off the protective shell around the viral chromosome, and take over the cell's machinery. Like a renegade information tape slipped into someone else's computer, the viral genetic instructions divert the system from its own built-in programming. The cellular machinery sets to work making new copies of the viral chromosome. Then it uses the viral instructions to make protein molecules to build shells for the new virus particles. The infected cell is forced to supply not only the machinery but all the molecular components and energy to make the new viruses. In return, the host cell is often destroyed as the virus particles proliferate. In an animal the infection and death of cells in various tissues result in diseases ranging from yellow fever, rabies, polio, and some cancers, to the common cold, measles, mumps, and chicken pox.

In the 1950s experiments by Joshua Lederberg, now president of Rockefeller University, and his co-workers revealed the possibility that we might be able to turn the table on viruses, exploiting their life-style to our advantage by using them to carry foreign genes into the cells they infect.

Viruses that infect bacteria—phages—do not always commandeer the cell's machinery and begin replicating themselves as soon as they invade. Instead they may lie low and take a rest for a few bacterial generations. They can snip apart the bacterial chromosome at a precise spot, splice in their own DNA, and integrate so completely that the host cannot tell the phage DNA from its own. When the bacteria copies its own DNA and divides, it copies the phage genes, too, and passes the dormant freeloader on to the next generation of bacteria. Then, generations later, when its host is threatened by chemicals or X-rays, or weakens and stops growing, the phage may become active again, snipping out its DNA, taking over the host cell, copying itself, and hitting the road.

However, the phage does not always do a precise job of excising itself from the bacterial chromosome. Occasionally, it snips off a chunk of bacterial DNA along with its own. The cellular machinery treats the whole string of DNA as though it were a phage, and the new viruses it turns out all carry these bacterial genes.

What Lederberg discovered was that when these phages with their load of pirated bacterial genes infect new bacteria, the genes go to work in the new host. The foreign genes brought in by the phage can even endow the host bacteria with new powers. The phage Lederberg worked with is called lambda. When lambda does a sloppy job of excising itself from the DNA of the common gut bacteria *E. coli*, it often picks up the gene that gives its host the ability to process lactose, milk sugar. Lederberg took lambda phages that had been infecting lactose-fermenting gut bacteria. Then he put them in a dish with related bacteria that lacked the ability to process that sugar. When the phages invaded the bacteria in the dish, they conferred on their new hosts the ability to ferment lactose. This type of gene transfer was dubbed "transduction."

(Phages are not the only viral pirates. As we have learned, researchers have discovered recently that the genes which allow certain animal tumor viruses to trigger malignant growth of the cells they invade are not viral genes at all. The viruses captured these oncogenes from normal cells.)

The possibilities for exploiting transduction were immediately apparent. Lederberg became one of the first outspoken advocates of eventual "genetic vaccines . . . viruses especially developed to carry correct genetic information to the body cells of patients with certain specific defects."

At the time, there were no techniques for isolating functioning genes, cloning them in large quantities, or splicing them onto viral vehicles for transport into cells. Scientists dreamed of developing mutant viruses with appetites for picking up specific genes of interest: "One can imagine farms of cultured human cells producing desired markers [genes] and delivering them up to viruses that can then profusely replicate them and infectiously transfer them to pregnant mothers, sick patients, and other unlucky beings adjudged by the family doctor to be deficient in something he knows about," Rollin D. Hotchkiss of Rockefeller University speculated.

There was another possibility, however, that did not involve loading genes onto viruses at all: In a few rare cases, a virus itself might possess a gene for some protein a patient lacked. This was the prospect that led biochemist and physician Stanfield Rogers to attempt in 1970 the first gene therapy on a human.

The saga of the first attempt at genetic therapy begins with rabbits. Early settlers in Kansas must have been startled by their first glimpse of wild cottontail rabbits scurrying across the prairies. The animals seemed to have horns projecting from their heads. But closer inspection of these "horned" rabbits showed they did not have horns, just warts. The warts—benign skin tumors called papillomas—are caused by a virus known as the Shope papilloma virus.

The Shope virus is what scientists call a "passenger" virus, an apparently harmless freeloader that does not kill the cells it infects. Only when it contacts the abraded skin of rabbits does it cause the abnormal cell growth that results in warts. In other mammals, including man, the virus produces no visible symptoms of infection. But in the late 1950s, Rogers began to find that the virus did have curious unseen impacts on the animals it infected.

Rogers was working with the Shope virus at the Oak Ridge National Laboratory in Tennessee, studying the process of tumor formation, when he discovered that the virus caused the skin cells of rabbits to make an enzyme called arginase, something they do not normally do. Further studies showed that the enzyme was different from the arginase produced normally in rabbit liver. The skin cells were apparently using information encoded in the viral gene, not their own switched-off arginase gene, to manufacture the enzyme.

When Rogers examined the blood of infected rabbits, he found that the enzyme that had been produced with the viral instructions caused a harmless but real impact on the animals: It significantly lowered their blood level of the amino acid arginine, one of the building blocks the body uses to make protein. The normal functions of the enzyme arginase are to break down excess arginine and to keep blood levels of the amino acid under control. Apparently the virus-coded enzyme, added to the rabbits' own, was overdoing the job, although not to a harmful degree.

Rogers began to wonder about all the people who had worked with the virus in laboratories over the years. Because exposed workers showed no obvious symptoms, it had been assumed the virus did not infect man. He took blood samples from these workers and found that about half of those who had worked with the Shope papilloma virus—even some who had not been in contact with the virus for decades—had blood arginine levels that were significantly lower than those of unexposed individuals. As in the rabbits, the lowered amino acid levels seemed to cause no harm. Rogers's studies convinced him that the cells of the lab workers, as well as those of the rabbits, were carrying out the virus's genetic instructions. Some critics, however, contended that the virus might simply be turning on or turning up the activity levels of the native arginase gene.

A report of the finding in *Newsweek* in 1967 noted that genetic diseases might someday be corrected "much the same way a mechanic fixes a car by replacing a burned-out spark plug." Rogers called himself "pathologically optimistic" in hoping the finding could be applied within five years.

But at the time, it was an interesting result with no apparent practical application: an ideal passenger virus all loaded up and ready to deliver a gene no one seemed to need. "It was clear that we had uncovered a therapeutic agent in search of a disease," Rogers wrote later. "How we wished then that it was carrying information for phenylalanine hydroxylase, as we might have been able to do something for phenylketonuria [PKU]."

But a potential use for the arginase gene was not long in coming. In 1969 Joshua Lederberg alerted Rogers to a report in a British medical journal. The report described the discovery in Germany of two little sisters with the opposite test results from the Shope-infected lab workers—extremely high blood levels of arginine. The condition was clearly a recessive hereditary disease: The girls' parents and two siblings had somewhat elevated levels of the amino acid, indicating they were carriers of the defective gene. But the two girls had inherited a double dose of the gene. The condition was dubbed argininemia, and the results were tragic.

"The older child—about five years old then—was having a really tough time walking across the room," Rogers recalled as he recounted his first visit to Cologne. "Both parents had to hold a hand to help her. She was obviously retarded and suffered convulsions. The two-year-old was in pretty good shape then. It didn't appear that anything was wrong with her."

But the blood of the younger child, and that of a third sister born about a year later, showed the same extreme excess of arginine found in the oldest girl. Progressive epilepsy, spasticity, and retardation seemed to be inevitable for them, too. The brain damage apparently was not a direct effect of either the lack of arginase or the buildup of arginine. But these imbalances threw off a more critical chemical cycle, one that rids the body of ammonia. The body normally handles ammonia by converting it to arginine. But since the girls already had elevated arginine levels, their systems were apparently inhibited from synthesizing more of the amino acid, and the excess ammonia built up to toxic levels.

Could the Shope virus do for the girls what it had apparently done for the infected lab workers—provide genetic instructions to make a product that would lower their blood arginine levels? If so, Rogers thought, the girls' ammonia intoxication might be reversed and the progress of the brain damage halted.

As a first step, he took some cells from the girls, grew them in a lab dish, and infected them with the virus. The results eventually showed that the cells produced the same apparently viral-coded arginase found in the rabbit skin cells. Rogers and the European physicians treating the children decided "with much trepidation but great hope" that the time was ripe for a first attempt at introducing useful new genes into human patients.

Rogers flew to Germany carrying a vial of Shope virus packed in ice. No one, then or now, knew how much virus would be needed to infect enough cells so that they would have a chance of correcting the disease. Because nothing like this experiment had been attempted before, the team proceeded with extreme caution in infecting the children—perhaps too much caution to give gene therapy a valid test.

"The dose of virus first used was about one-twentieth of that which we had previously found harmless to mice," Rogers reported later. "As might be expected from this tiny dose, no effect whatever was found either in the condition of the children or in their arginine or blood ammonia levels."

The oldest girl continued to deteriorate, and a year after the first injection she was given a second, rabbit-sized dose. "A short time thereafter the blood arginine level dropped transiently, but it subsequently returned to its original high levels. Thereafter, however, the child gradually improved. It is not certain, however, whether this was related to the virus or was an effect of the low-protein diet, which was known to lower ammonia levels," Rogers wrote.

When the family's third argininemic child was born, the team decided to try to head off the tragic effects of the disease with early therapy. Rogers shipped more purified virus to Germany. But inoculation of the baby was delayed for several months, and the unstable virus preparation "went to hell." It was later found to have no infectious power left. The inoculation had no effect on the child's blood arginine or ammonia levels.

By that time—1971—a tide of professional criticism had swept over Rogers, suggesting his work was premature, and further tests in humans were halted. His funding withered away and he turned to working with plant viruses. Rogers, now at the University of Tennessee, has remained convinced that the experiments were morally correct. "Although these results were at best disappointing, it still seems to us that the chance to prevent progressive deterioration in these children's condition was the only ethical route to take and that, should other children be found to have this disease, they should be so treated," he wrote later.

Whether Rogers's attempt to treat human patients was premature or not, it was clear to him and to others that there was a limited usefulness to the system he had discovered. Finding harmless passenger viruses conveniently loaded with genes needed to treat other, more common hereditary diseases seemed like a long shot. So was finding viruses that had pirated genes of use in treating human disease.

In 1971 Carl Merril of the National Institutes of Health generated worldwide headlines when he reported infecting human cells with lambda phage carrying the bacterial gene for processing milk sugar—the same phage Lederberg had watched as it transferred the gene between bacteria.

The human cells Merril and his co-workers used had been taken from the skin of a patient with the inherited disease galactosemia, which leaves individuals unable to handle milk sugar. If a baby with this disorder isn't taken off milk and milk products, the result can be retardation and death. Merril's experiments indicated human cells could use the bacterial gene to produce a sugar-processing enzyme similar to the one they lacked. Many other labs tried to duplicate the experiments and failed, however. The claims remained controversial for years, as did the very notion of trying to repair a defective human cell.

For those who did see a bright future for gene transplants, the problem remained of finding suitable viruses loaded with genes of interest. What was needed was a way to load desired genes onto viruses—methods known today as gene splicing or recombinant DNA techniques. Even as Rogers and Merril were drawing fire, those new techniques that were to revolutionize molecular biology and transform gene therapy from bizarre heroics into a realistic goal were already being worked on in research labs on the West Coast.

15. The Monkey Virus

> Work on recombinant DNA thus recalls many nightmares.
> It has the smell of forbidden knowledge. It calls up old
> myths in which mortals were harshly punished for having
> stolen a power exclusively reserved to the gods. Especially
> outrageous is the demonstration that it is so easy to fool
> about with the substance that is at the very basis of all life
> on this planet.
> —FRANÇOIS JACOB, French biochemist, 1982

Around 1970 it became obvious that isolated genes would be available sooner than anyone had imagined, opening the way to an understanding of the precise molecular anatomy of heredity. But scientists realized that learning the sequence or structure of genes would not tell them why they do what they do when they do it: how genes for insulin respond to blood-sugar levels; why growth-hormone genes are turned on only in the pituitary. Some method was needed for putting genes back into a cellular environment in a controlled way, in a place or condition where their activities could be watched.

Molecular biologists began to tinker with various vehicles and systems that would let them insert isolated genes into alien cells. One first, overly optimistic hope was to find animal viruses that could load and transport the desired genetic cargo on their own, the way phages capture bacterial genes. But when that prospect dimmed, researchers decided to do the loading themselves. They began to design ways to splice foreign DNA sequences into the genetic material of viruses or bacterial plasmids. Today every schoolchild recognizes their invention as the basic technique of recombinant DNA.

Stanford University biochemist Paul Berg focused his attention on a simple virus already familiar in cancer research labs—simian virus 40 or SV40. Its name reflects its preferred target. In nature, SV40 is interested only in infecting monkeys. Once inside a monkey cell, an SV40 particle strips off the protein shell that covers its minichromosome (5,243 nucleotide molecules long, containing only 5 to 7 genes), commandeers the cell's machinery, and makes hundreds of thousands of copies of itself. The crowd of viruses finally bursts the host cell and the parasites move on to infect new cells.

Berg and his colleagues began knocking out little pieces of SV40's DNA, making substitutions, additions, and deletions to determine the function of each piece of genetic material. If foreign genes were to function, they had to be inserted into the right spot along the stretch of viral DNA so that the viral signals would be able to activate them in the infected cell. Room for the foreign genes had to be made by deleting pieces of the virus's own DNA. The researchers found that this gutting crippled the virus so that it could not multiply on its own. This handicap could be overcome, however, if the target cell was also infected by a second, "helper" virus. The helper's genes coded for the necessary protein products the defective SV40 virus could not supply itself.

By 1971 Berg's group had taken the first step toward a gene transfer system, demonstrating that they could hook a long strand of DNA from lambda phage and the *E. coli* bacteria onto the genetic material of SV40 to form hybrid molecules. It was the first time humans had ever recombined DNA from two different organisms, and it led to a 1980 Nobel Prize for Berg.

Berg's team had planned to take the hybrid molecule and insert it into *E. coli*. The idea was to let the bacteria grow and divide, copying and passing along the hybrid DNA with each generation, and thus producing large quantities of identical recombinant molecules—the procedure we know today as molecular cloning. But Berg backed off when other scientists began to suggest it might be hazardous to introduce novel viral instructions into a bacterium that thrives in the human gut. Reaction to this proposed experiment helped initiate the years of controversy over the safety of recombinant DNA.

(In 1973, shortly after Berg halted his work with the recombinant virus, Stanley Cohen, another Stanford scientist, and Herbert Boyer, a researcher at the University of California in San Francisco, actually became the first to clone a recombinant molecule in bacteria. Instead of a viral vehicle, they used a bacterial plasmid to carry the foreign genes into the bacteria. The plasmid carried a gene for resistance to streptomycin, further fueling the safety concerns about the new powers scientists might be conferring on common bugs.)

But making monsters was not the object. Recombining viral

and foreign genes was only part of the goal, and bacteria only one possible host. If SV40 could invade monkey cells and make hundreds of thousands of copies of its own DNA, perhaps it could be tricked into doing the same for a foreign gene if the gene were tacked onto the virus in the right place. The procedure would be a useful way to study certain gene functions, by multiplying a gene or pieces of a gene and exaggerating its activities in an alien animal cell where it could be easily observed.

Berg and Richard Mulligan, his student at the time, attempted to put together a hybrid viral and foreign gene, infect an animal cell with it, and get the viral signals to turn on and express the foreign gene. The experiments were performed in a high-containment lab, which was required by the federal safety code enacted as a result of the concerns raised by the work being done by Berg and others.

"What we were asking then was, could we just churn out lots of a protein?" recalls Mulligan, now at MIT. "The SV40 transcriptional signals are very strong, so that late in the virus infection, lots and lots of particular viral proteins are synthesized. And the question was, could you just substitute viral protein-coding regions with foreign protein-coding sequences and fool the virus into thinking it was making a viral product?"

The answer was yes, and the results made headlines in 1978. The researchers took rabbit genes for hemoglobin components and hooked them up to SV40 virus. They used these hybrids to infect kidney cells from an African green monkey. Monkey kidney cells do not normally produce any blood components, even monkey ones. And yet the infected cells began churning out the proteins needed to make rabbit hemoglobin. The credit went to SV40. The virus, thinking the rabbit gene was one of its own, used the monkey cell machinery to manufacture large quantities of the protein coded for in the rabbit instructions.

The experiment marked the first time recombinant DNA techniques had been used to transfer genes from one mammalian species into the cells of another. Later the Stanford team spliced genes from fruit flies and then from mice into the SV40 DNA, and infected monkey cells with these recombinants. The monkey cells turned out fruit fly and mice proteins as efficiently as they had made rabbit products.

The system was obviously highly efficient, but it had serious limitations. Most important, it eventually killed the infected cells.

That made the method unsuitable for putting genes into whole animals or treating human disease as well as for many basic genetic experiments Berg was interested in. A second limitation was the architecture of the virus particle itself. The capsule, which is necessary for the virus to remain infectious, can hold only a fixed amount of DNA. This greatly restricts the size of the foreign gene that can be spliced into the virus. And finally the virus can reproduce only in monkey cells, not in human, mice, or other animal cells.

In the cells of other creatures, SV40 cannot complete its life cycle or kill the cells. But occasionally it can cause another problem for the host animal. It can turn infected cells into malignant tumor cells. It was this trait that had made the virus an attractive object of study for cancer researchers. In mice and other animals, only part of the SV40 genes get turned on. These signals are not adequate for the virus to take over the cell's machinery and reproduce itself. But they are enough to foul up the system and trigger the uncontrolled and invasive cell growth known as cancer.

Engineering a viral invasion had proved to be an efficient but somewhat destructive way of introducing new genes into cells. So the Stanford team turned its attention away from using viruses as vehicles. The bits of viral signals they had so painstakingly located among the SV40 genes could still be snipped out and hooked up to foreign genes to induce them to work in new settings. But without the genes that give the virus its destructive and infectious powers, these hybrid molecules could not invade cells as viruses do. Another way had to be found to get them into cells.

In the 1970s researchers had begun to copy another gene transfer system found in nature. It was a way of coaxing cells into voluntarily taking up new DNA—a means of gentle persuasion rather than viral invasion called DNA transformation.

16. Soaking Up Naked DNA

> One of the lessons of history is that almost everything one
> can imagine possible will in fact be done, if it is thought
> desirable; what we cannot predict is what people are going
> to think desirable.
>
> —P. B. MEDAWAR, 1963

Leaning on the chicken-wire fence of his duck pen one spring day in 1957, Father Leroy had quite a start. "It is wonderful to be close to the workings of creation," the Jesuit later exclaimed to the French press as he told of that day. His ten-month-old ducks seemed to be going through an amazing metamorphosis, and he and his colleague were convinced they knew why: They had been injecting extracts of DNA into the ducklings' body cavities since they were eight days old.

"The French press has been full of praise, speculation and wonderment, not unmixed with uneasiness," *Time* magazine reported, noting that DNA is "believed to be concerned with heredity." Father Leroy, "a Jesuit priest-biologist and refugee from Red China," and Professor Jacques Benoît of the Collège de France, who had "studied the reproduction of ducks for 20 years," were partners in the adventure. The pair had started the previous summer with a dozen hatchlings of the Pekin breed—large, creamy-white, orange-billed ducks—"bought from a reliable dealer." The injected DNA came from the genitals of Khaki-Campbell ducks—smaller brown birds with greenish-black beaks.

The first change Father Leroy noticed came the following March: The ducklings' bills were turning greenish-black at the base. As time went on, the two researchers noticed more changes that seemed to set their flock apart from its Pekin parents: the softness and color of their feathers, the angle of their necks, their placid temperaments. The two proclaimed a new type of duck— *blanche-neige*, or snow-white.

While Benoît and Father Leroy were assuring the press that similar injections for treating human diseases were a long way off, French scientists questioned the validity of their claims that exposure to foreign DNA could have made such major changes in the animals.

Time seems to have proved the critics right, and history has bypassed the metamorphosed ducks. But in 1957 scientists had good reason to hope that some animal cells—if not whole animals—might be significantly and permanently changed by contact with naked DNA. The process is known as DNA transformation. (It has no relationship to the malignant "transformation" a cell undergoes when it turns cancerous.)

Transformation was first observed half a century ago in bacteria. In fact, it was transformation that first led scientists to identify DNA as the material basis of heredity. As researchers bathed bacteria in DNA extracts and watched them being transformed from one genetic type to another, it became clear that genes must be made of DNA. Not long after they identified the stuff of heredity, scientists began trying to mimic nature's methods for moving it about.

The story began in the 1920s when a London physician named Frederick Griffith injected mice with two forms of a pneumonia-causing germ—one living and one dead. The living bacterium was a mutant form of *Streptococcus pneumoniae* that had lost its ability to cause the disease; these bacteria were known as the R form. The second germ was a virulent strain of the same bacterium, called the S form. Griffith killed a batch of these deadly S forms with heat and injected them into the same mice. Since the only living bacteria the mice received were harmless, no infections should have occurred.

Instead the result was startling: Many of the mice died. In the blood of these dead animals Griffith found living, virulent S-form bacteria. What had occurred was no miraculous resurrection. Something from those dead S forms he injected had transformed their harmless R-form cousins into killers. That unknown material had transferred a specific ability from the S forms to the R forms, the ability to cloak themselves in a capsule of complex sugars. It was this capacity which the mutant R forms had lost, the trait they needed to be normal, pathogenic S forms again. Once the transformation of the R forms occurred, it was permanent. Griffith cultured generation after generation of these transformed microbes and got only sugarcoated S forms.

It was clear that whatever had slipped through the cell walls of the R forms and changed the microbes into S forms must be the stuff of heredity—the genes. But at that time no one knew what genes were made of. Scientific opinion leaned toward protein rather than DNA.

American microbiologist Oswald T. Avery of the Rockefeller Institute for Medical Research in New York City set to work to find and identify this transforming principle. He took purified extracts from S-form microbes and put these in glass dishes with R forms. The R forms were changed into S forms just as Griffith's had been. Avery then studied these extracts exhaustively. In 1944 he published his findings: The substance responsible for making permanent, inheritable changes in the microbes was DNA.

As soon as Avery's work was published, scientists began to speculate that human genetic information might someday be altered in much the same way. But there were basic questions to be answered first. Chiefly, could the complex, nucleated cells of higher animals simply soak up and use genes they ran across the way bacteria did?

In the same year that Benoît and Father Leroy began injecting their ducklings, a team at Rockefeller University was launching a less ambitious but more meticulous test: bathing skin cells from white mice embryos in DNA taken from black mice. The researchers inserted the treated cells under the skin of other young white mice, hoping one of these growing clumps of foreign cells would produce a dark hair, indicating it had picked up a pigment gene from the black mouse DNA.

"After we had exhausted all the arts we had acquired or thought of in telling bacteria information their own mothers had never told them, we found that we had been able in one whole year to test and follow up only some 20 million mouse cells, but not one seemed to acquire the pigment marker [gene]," Rollin D. Hotchkiss—who had worked with Avery to identify DNA as the transforming agent—wrote with wry humor nine years later. "This was enough to tell us that DNA transformation of mammalian cells was not going to be easy, and we went back to the more malleable bacteria." Yet Hotchkiss was one of the first to prophesy the coming of genetic engineering.

The prospects for transformation of animal cells remained dim until only a decade ago. Some of the 1960s experiments did show that isolated chromosomes could penetrate animal cells, but the cells did not appear to make use of the new DNA.

The picture changed in 1973. O. Wesley McBride of the National Cancer Institute and his colleague Harvey Ozer bathed mouse cells with a solution containing DNA from Chinese hamster cells. The trick was to spot which mouse cells, if any, had absorbed hamster genes and actually put them to work. For this, the researchers used a technique that is still crucial to most gene transfer procedures today—selection. They simply made it impossible for the mouse cells that did not take in certain hamster genes to survive.

The mouse cells McBride and Ozer used were mutants. They lacked the ability to make the critical enzyme HGPRT. As we saw in Chapter 10, these mutant mouse cells cannot live when their culture dishes are filled with the specially designed nutrient broth called HAT medium. The hamster chromosomes carried the gene needed to make HGPRT, and this gene along with the rest of the hamster DNA was offered to the mouse cells before they were subjected to HAT.

The team found that a few, a very few—at most, one in a million—mouse cells had been transformed by the hamster genes just as Frederick Griffith's microbes had been. They were able to grow and form colonies in the hostile HAT medium. Apparently those few cells had engulfed hamster chromosomes—a process called endocytosis—and bits of this foreign DNA had escaped the cells' digestive enzymes and survived to function. Normally when a cell bumps into an interesting particle or molecule, the cell membrane folds around it and forms a vesicle, drawing it inside. If the vesicle fuses with a lysosome, one of the cell's digestive sacs, the enzymes inside the sac can degrade the foreign particle. This is how the cell ingests and digests nutrients and protects itself from bacteria and other perturbing things.

(A cell's defenses against intruders are formidable enough to make transformation a rare event. The chemical hazards that faced unprotected DNA in the body cavities of Father Leroy's ducks make it doubly unlikely that many genes found refuge inside nearby cells. And even if all the cells of the visceral lining had received new genetic instructions, it is questionable whether those cells could have had a strong impact on the bearing, disposition, and plumage of the birds.)

Other successes followed, and transformation techniques were improved. But offering cells a whole chromosome full of genes is haphazard and imprecise, like cell fusion. A way was needed to get specific genes of interest into cells.

In 1977 a group at Columbia University College of Physicians and Surgeons—Richard Axel, Saul Silverstein, and Michael Wigler—began to use the transformation technique to move isolated genes instead of whole chromosomes into cells. They started with a viral gene for an enzyme called thymidine kinase, or TK. TK is another enzyme which, like HGPRT, is necessary for survival in HAT medium.

Using modifications of techniques that had been tried in the mid-1970s, the Columbia team developed the transformation methods that remain in wide use today. With them, scientists can now transform as many as one of every ten thousand cells they expose to foreign genes.

Instead of pouring a clear, watery solution containing DNA directly onto cells in a lab dish, as earlier researchers did, scientists now mix the purified genes into another liquid, a calcium phosphate solution. In a half hour, the liquid becomes slightly cloudy, indicating the calcium is precipitating out of solution with the genes. The precipitate looks roughly like microscopic salt. A few drops containing the precipitate are added to the watery nutrient medium in which the cells that will be transformed are growing, and the dish is incubated overnight at 37 degrees. For reasons not fully understood, the calcium helps more of the genes become engulfed by cells, survive digestion in the cytoplasm, cross into the relative safety of the nucleus, and insert themselves into the host's chromosomes.

After incubation with the genes, the cells are subjected to some hostile selection process to weed out the vast majority that fail to produce the product coded by the foreign gene. Axel's group used mutant mouse cells that lacked a TK gene and could not survive in HAT unless they had successfully made use of the viral TK gene offered to them.

This sort of selection, however, limited scientists to working with cells that lacked something and genes that could provide what they needed. Some of the genes that scientists were most interested in working with, including many involved in human hereditary diseases, function only in specialized tissues and provide no survival value for the cell itself. For instance, the genes for globin, one of the components of hemoglobin, are important to an animal that depends on blood to get oxygen and nutrients to its cells. But globin genes do nothing to help a cell survive in a glass dish. Could an animal cell be made to take in and keep any gene, even one that it didn't need?

The answer was yes, and the cue came from studies in bacteria. Very few bacterial cells are able or willing to take up and use a foreign gene, but those that are seem to have a more-the-merrier attitude. They are likely to acquire more than one new gene, a process called "co-transformation." Axel's group set out to learn whether it occurs in animal cells, too. They exposed mouse cells to genes for human growth hormone along with the virus TK genes, then put the cells in HAT. Of the survivors—i.e., those that had successfully taken up TK genes—80 percent also took up the human genes. Later it became evident that all the foreign genes a cell takes in—usually multiple copies of whatever is offered—link up in a chain and are inserted as a unit somewhere at random along the host chromosomes.

At almost the same time that co-transformation was proving successful in animal cells, Stanford biochemist Paul Berg was using viruses to transfer specific mammalian genes into the cells of other mammalian species for the first time. Shortly afterward, scientists around the world began to use co-transformation to do what Berg and others were doing with SV40—coaxing mouse cells into taking up rabbit globin genes, chicken ovalbumin (egg albumin) genes, human globin, growth hormone, interferon genes, and others.

Next, Berg and Richard Mulligan set out to combine the benefits of both systems. They would strip away all the genes the monkey virus needed to infect a cell, replicate and encapsulate itself, and kill its host. But they would keep the viral "promoters" and other signals that had proved so effective in getting foreign genes activated and working in infected cells. What they would be left with when they hooked the viral signals to a foreign gene would be a naked molecule of recombinant DNA, not an infectious virus. However, using the technique of transformation, the researchers should be able to insert the hybrid DNA into a cell and get the foreign gene to work proficiently.

"By then the viral genome was so well characterized that we

could just slap together hybrid genes at will using pieces of viral signals that were very strong," Mulligan says.

The researchers decided to link a bacterial gene to the viral signals to see if they could correct an enzyme deficiency in human cells. The gene they picked is the bacterial equivalent of HGPRT in man, called XGPRT. The target cells, taken from a human patient with Lesch-Nyhan syndrome, lacked an effective HGPRT gene and could not cope with HAT medium. The team hooked the viral signals to the bacterial XGPRT gene and transformed the human cells with it. The enzyme deficiency was corrected. The cells of the Lesch-Nyhan patient could survive in HAT.

"It was a big step for us because it was really the first time one demonstrated you could actually express a bacterial gene in animal cells," Mulligan says. "It blew aside all these myths that maybe there's something special about animal cell protein-coding sequences."

Berg quickly quashed any suggestions that his team had come up with a method of gene therapy, noting that transforming the genetic information of a few cells in a glass dish was worlds away from putting new genes into the one hundred trillion HGPRT-deficient cells of a human patient. The efficient functioning of the transferred gene had been achieved with DNA signals from an animal tumor virus—something we might not want to put into human cells, he suggested.

The team's chief goal had been to design a new selection method to allow the transfer of genes into normal cells instead of into mutant animal cells. If bacterial XGPRT genes could be made to work in animal cells, they would confer on them a new ability animals ordinarily do not have or need—the ability to process a compound called xanthine. Then by manipulating the environment inside a culture dish, scientists could make it necessary for the cells to handle xanthine to survive. Normal cells that hadn't acquired the XGPRT gene—along with any other genes the researchers were interested in co-transforming—would not survive in the dish.

Other labs were developing new selection schemes, too, including the use of genes that conferred drug resistance on transformed cells. The net effect was to expand the range of cells into which scientists could transfer new genes. DNA transformation techniques were soon being used worldwide as a basic tool of molecular biologists, making possible the isolation of important genes, including human cancer genes, and providing a way to study their regulation and control.

If Berg and other basic scientists thought gene transfer was still in too primitive a stage to start thinking about applying it to human therapy, some clinical researchers did not. As the decade of the 1970s drew to a close, many were optimistic that the remaining hurdles in the rapidly expanding field would fall quickly, and a few thought it was time to begin gearing up for moving new genes into human patients.

17. New Genes for Mice and Men

Paradoxical forces exist that tend to spur us to action.
People clamor for the fruits of research to be brought from
the laboratory into the public domain. Public funds are
spent and, for financial support to continue, practical
applications are expected in the near future. As a result,
premature applications are likely to be attempted. The
public wants cures and prevention of disease; yet, for some
of the most serious problems the basic knowledge that
would enable us to "deliver the goods" is lacking.
　　　　—ARNO G. MOTULSKY,
　　　　　　University of Washington geneticist, 1974

In 1980 gene transplants moved from lab dishes to mice to human
patients with dizzying speed and generated a barrage of newspaper headlines and network television reports. Transformation
techniques finally seemed about to achieve what Father Leroy
had hoped for in his ducks—a tangible change in some inherited
trait of a living creature. But many molecular biologists, still trying to get transferred genes to work normally in a lab dish, feared
that the work had gone too far too fast with questionable success.

　One of the frightening ironies cancer patients face is that the
treatment often seems more brutal than the disease itself. Researchers have found no way to target lethal drugs exclusively to
tumor cells and avoid some damage to normal body cells. But
there is one distinguishing trait of cancer cells that can be ex-
ploited—they grow and divide much more rapidly than most normal tissues. Powerful anticancer drugs like methotrexate have
their greatest impact on these rapidly dividing cells. Unfortunately, some normal cells also divide rapidly, and the drugs take
their toll. Bone marrow, which contains the body's blood-forming
cells, is especially vulnerable to methotrexate. The drug can de-

press the production of red blood cells and leave patients dangerously anemic.

Some cells, however, can overcome an assault by methotrexate and develop a genetic resistance to the drug, not a trait physicians want to see in cancer cells. But if this resistance could be conferred on the bone marrow cells of cancer sufferers, doctors would be able to administer larger doses of tumor-killing drugs without harming the patients.

When Martin Cline, a physician specializing in blood disorders, and his team at the University of California at Los Angeles set out to transplant genes into living mice, they chose to transfer genetic protection against methotrexate into mouse bone marrow.

Methotrexate kills cells by preventing them from utilizing an essential vitamin, folic acid. It does this by blocking a necessary enzyme, dihydrofolate reductase, or DHFR. But some cells are able to pull an end run on the drug and evade death. They make multiple copies of their DHFR gene and begin producing more of the enzyme than the drug can block. It was extra copies of the DHFR gene that Cline wanted to transfer into mouse cells.

Using the transformation techniques developed by Richard Axel's group only a few years before, Cline took mouse marrow cells and bathed them in mouse DNA enriched with extra DHFR genes. Then he added a new twist to the method. Instead of putting the cells in a culture dish and subjecting them to selection—exposing them to methotrexate so that only those that had taken in extra DHFR genes would survive—Cline injected all the treated marrow cells directly back into mice whose own marrow had been destroyed by radiation. Then he subjected the living animals themselves to selection.

Daily doses of methotrexate were given to the mice, making it tough for the bone marrow cells that still had only one DHFR gene to proliferate. The cells with multiple protective genes flourished. The animals developed increased resistance to the damaging effects of the drug.

When the work was announced in April 1980, the press reported it as "revolutionary." But fellow researchers were skeptical. The DHFR gene Cline offered the mouse cells was indistinguishable from the one they already had. If DHFR production had increased under the assault of methotrexate, was it because the cells were using transplanted genes? Or had some of the marrow cells simply mutated and multiplied their own DHFR genes?

Cline reported another similar set of live mouse experiments the same month, this one using the viral TK gene. Again, critical colleagues, examining the blots and bands on X-ray film by which genetic engineers track fragments of DNA, saw no firm proof that viral genes had been successfully integrated into the mouse DNA. More than three years later, no one else had been able to repeat these animal experiments successfully, a process that is necessary in science before results are accepted as valid.

Cline, a world authority on bone marrow cells but a relative newcomer to molecular biology, was undaunted by the hesitant response of his nonphysician colleagues. In scientific articles, press releases, and interviews with journalists, he stressed the promise these gene transfer techniques seemed to hold for eventually treating patients with hereditary disease and easing the side effects of cancer treatment. The applications to human cancer patients might come in "three to five years," he told a reporter then. And it was "theoretically possible," he wrote, to use this gene transplant method to treat blood disorders such as sickle cell anemia and the thalassemias.

But Cline, aggressive and self-assured, was not planning to wait five years or watch cautiously while someone else made the first move toward gene therapy in humans. For almost a year he had been seeking permission from UCLA review committees to try transplanting new genes into patients with sickle cell anemia. Similar requests were pending at hospitals in Israel and Italy, where Cline wanted to try the therapy in thalassemia patients. The UCLA reviewers were clearly skeptical of the request. The human globin genes that would be used had never been made to function efficiently in transformed cells, and Cline himself had failed at getting the human genes to work in mice. But he was convinced the time was right, and less than four months after his work in mice was announced, Cline attempted the first transplant of human genes into a human patient.

Hemoglobin is the oxygen-binding protein that colors our blood red and keeps all our one hundred trillion cells breathing. Life for hemoglobin-filled red blood cells is brief and hectic. Surging through the tens of thousands of miles of blood vessels in the body, delivering oxygen and picking up waste carbon dioxide, the cells collapse from wear and tear and die in about four months. This, in turn, makes life a rat race for our bone marrow, the soft, spongy tissues found in the cavities of our larger bones. To keep us from getting anemic, the marrow cells must manufacture about two and a half million replacement red blood cells every second.

Real trouble occurs when one of the genes that provides blueprints for the various hemoglobin components drops the ball. The gene may work inefficiently, or not at all, or pass along botched instructions. The result at the level of the bone marrow is a defective product, an imperfect red blood cell. For the person in whom the genes are housed, the result can be a devastating blood disease.

As we have seen, a group of hereditary blood disorders, the thalassemias, is prevalent among the peoples of the Mediterranean, Middle East, and Far East. One form, beta-thalassemia or Cooley's anemia, is caused by a defect in the gene for one of the two main protein components of hemoglobin—beta-globin. The result of the defect is production of red blood cells that are brittle and shatter easily in the rough-and-tumble raceway of the circulatory system.

By the time Cline's animal experiments were announced in the United States, a young woman in Jerusalem had been struggling with beta-thalassemia for twenty-one years without hope of a cure. With her fragile red blood cells breaking down at a life-threatening rate, her body had tried to compensate by producing excessive quantities of bone marrow. The extra marrow had deformed her bones, causing disfiguring fractures and distorting her face into oversized gargoyle-like features. She walked painfully with a limp. Her body—especially her heart—was littered with accumulations of iron, the debris of billions of shattered blood cells. Frequent blood transfusions added to the iron problem, but she needed them to stay alive. Chronic heart failure sent her to the hospital frequently.

The young woman wanted to spend what remained of her life teaching handicapped children—a brief prospect, since few beta-thalassemia sufferers survive their twenties.

In Italy at the same time, a sixteen-year-old girl from Turin had not yet entered puberty. She resembled a twelve-year-old in stature. Like the Israeli woman, the teenager had been struggling all her life with the effects of beta-thalassemia. Blood transfusions begun at an early age had spared her from painful bone deformities, but as a consequence her heart was already failing under the effects of iron overload and her development had been retarded. Each night she slept with a needle inserted beneath her skin as an automated infusion system slowly supplied her body with a drug to help excrete some of the excess iron. It was, however, only a stopgap effort.

Martin Cline was still waiting for permission to do the human experiments in all three countries when he flew to Naples and then on to Jerusalem to meet with colleagues there in the summer of 1980. He had carried along with him a supply of isolated human beta-globin genes as well as viral TK genes, planning, he said later, to do some experiments in cultured cells while he was abroad.

Early in the morning of July 10, officials at Hadassah Hospital in Jerusalem notified Cline that he had been granted permission to attempt human gene therapy. By 9:00 A.M. the same day, his Israeli patient, the twenty-one-year-old beta-thalassemia sufferer, had arrived at the hospital to prepare for the procedure. What followed must have seemed rather mild to her compared with the other medical procedures she had endured throughout her life.

That morning her hip was numbed with local anaesthetic, her hipbone pierced with a long needle, and about an ounce of bone marrow drawn out. The marrow was placed in a lab dish. Then both human globin genes and viral TK genes, precipitated with calcium phosphate, were added. (Cline believed the viral genes might give the cells a proliferative edge on the untransformed marrow cells.) While her bone marrow cells were incubating with the new genes, the patient's lower legs were given three hundred rads of radiation—a mild dose—to kill the marrow there and create a niche for the returning marrow, a standard practice in marrow transplants.

That afternoon, five hours after they had been removed, the cells were reinfused into the patient's bloodstream to circulate until they homed in on a niche in the marrow compartment. Of the 500 million cells, perhaps 5,000 to 50,000 might have been expected to pick up new genes. But only a handful could possibly have been stem cells, primitive units capable of proliferating and repopulating the marrow.

The next day the whole process was repeated, except for the radiation. A few days later, Cline flew to Italy and repeated the procedure on the teenager from Turin at the University Polyclinic in Naples.

Cline said later that these gene transplants would have been only the first in a series of transfers in these two patients, part of a study that he thought would span a decade. The goal of the initial infusions was only to see whether genes could be gotten into a human in this manner, and whether it could be done without harmful effects. Efficient functioning and control of the new genes would be worked on down the line. He felt that the chances the first transplants would actually help the two patients had been "exceedingly remote" from the beginning.

Like the effort a decade earlier to treat retarded patients in Germany with viral genetic information, however, Cline's attempt at gene therapy was cut short by controversy. The same week Cline finished the gene transfer in his Italian patient, a UCLA committee back in Los Angeles rejected his request to try the procedure in sickle cell anemia patients. The verdict: More animal studies were needed.

Word of the human experiments spread quickly through the scientific grapevine, and controversy was already boiling around Cline when the story hit the newspapers in October of that year. "Human Engineering: Pioneer Genetic Implants Revealed," the *Los Angeles Times* headlined its page one lead story. "Genetic engineering has been used on humans for the first time . . . ," the Associated Press reported around the world.

Many molecular biologists were reluctant to be too openly critical of Cline and risk starting a public controversy that might set back the real promise of gene therapy. The sometimes acrimonious public debates over the safety of recombinant DNA in the mid-1970s were fresh in their memories. But the lab scientists who work most closely with gene transfer were appalled and angry. Some told reporters that the work was "scientifically premature," a "shot in the dark," and that Cline had used "scientifically poor judgment" in moving into humans so quickly.

"The experiment was destined to failure," molecular biologist Thomas Maniatis, now of Harvard, told *Discover* magazine that fall. Maniatis had supplied Cline with the beta-globin genes, not knowing he planned to use them in humans.

"There is simply no scientific basis for expecting this experiment to work in people," Richard Axel, whose team developed the transfer technique, told the journal *Science*. "Cline has done this experiment in a mouse and, as I understand it, it didn't work. He has made a great conceptual leap from the failure in a model system to trying it in humans."

But physicians who work with dying patients, including the Italian and Israeli physicians whose patients Cline had treated, were more understanding of his decision and motives. "I'm a doctor who is trying to do the best I can for my patient," Eliezer Rachmilewitz of Hadassah Hospital told *Discover*. "This girl is at the end of the line. This is her only hope."

"When do you consider animal experiments adequate?" Cline asked a *Los Angeles Times* reporter, noting that it would take years for a full understanding of gene transfer and function to emerge from the labs. "Here's a patient who has a life-threatening disease with a limited life expectancy and no options with modern treatment. Is now the time to try an experimental treatment?" He firmly believed then, and has continued to affirm, that such decisions should be up to individual physicians and their patients. The treatment's potential hazards—that an out-of-control gene might produce too much of its product, or drift into a cell where it didn't belong, or bump a normal gene or a control signal out of place and possibly cause cancer—do not seem likely enough to occur to be a serious consideration in patients with such a limited life expectancy, he said.

Fellow clinician Theodore Friedmann of the University of California at San Diego told a congressional subcommittee hearing in 1982 that physicians in Cline's position may recognize "obligations strong enough to override the prematurity of the science.

"I think medicine often goes to rather extreme lengths in the face of terminal or untreatable illness to use untried and unproven new tools," he said. "I think much of what we come to recognize today as very effective therapy for cancer began, and even now continues to develop, as a result of these very unlikely shots in the dark. I am not at all convinced that medicine is or can be held to the same standards as less applied science. Ideally, it would be expected to be no less rigorous, and perhaps even more rigorous since human lives and welfare hang in the balance. I think that it is difficult to know when to accept what little knowledge is available and accept the risks of the early applications of imperfect knowledge to problems of human health. One doesn't always know when to accept one's ignorance or imperfect knowledge and proceed."

As for the young women who received the first gene transplants, apparently they were neither harmed nor helped. Limited analyses of their blood and marrow samples gave no indication that any new genes were working to allow the production of normal hemoglobin. Their fate is still sealed in their own genetic inheritance.

Reflecting on the experiments a year later, Cline remained somewhat dazed by the storm that had overtaken him. UCLA had accepted his resignation as chief of hematology-oncology, although he remained a tenured professor until 1983 when he moved to the City of Hope Research Institute in nearby Duarte. The National Institutes of Health, the federal agency that controls most medical research money in the United States, had found him guilty of violating federal guidelines for both human experimentation and recombinant DNA research. (Cline had broken the deadlock in the Israeli committees by changing the procedure he planned to use, eliminating any "recombinant" molecules and asking to transfer only purified genes. But at the last minute, without informing the Hadassah officials, he deviated from this, using genes that were linked together as hybrids. It was a formality, since genes transferred together tend to link up anyway, but Cline admits the move was "poor judgment" that he "greatly regrets." The NIH also ruled that his work abroad should have been approved first by committees at UCLA.) As a result, he lost $190,000 in grants, about half his federal funding, and extra layers of red tape were applied to his future work in the field. Two years later he still had received no new federal funding.

The punishment was harsh, but it reflected the public sensitivity toward anything that looked like less-than-well-thought-out genetic tinkering. The NIH appeals board later determined the penalty to be "clearly reasonable in relation to appropriate institutional concern about activities in an awesome and potentially dangerous area of science."

Whatever Cline regrets, it is not his faith in the eventual promise of gene therapy. "If it works, and I think it will, it would be analogous to bringing new treatment to untreatable conditions," he said in 1981. "This has happened a number of times in medicine—the development of vaccines, antibiotics. All allow you to treat things you couldn't conceive of treating before. I think most people agree that gene therapy is a real possibility in the future. No one I know of has said, 'No, you won't be able to replace genes.' The criticism has come in the timing and in how much we must know before we proceed."

When Cline appeared before the same congressional hearing at which Friedmann testified in 1982, he was asked whether he thought "another effort to treat humans suffering from blood disorders should be undertaken."

"Yes, I do," he replied.

Representative Albert Gore, Jr., challenged Cline's observation that the news media coverage of the controversy over his gene therapy attempts had helped to erode public confidence in science. "There is a growing feeling on the part of many that our reach is exceeding our grasp in too many instances and we are acquiring powerful new tools and not exercising the care and wisdom necessary to wield them carefully, and I think that the responsibility of individual scientists in this profession is going to grow, must grow if the public's confidence in science is to be maintained as it must be," Gore stated.

"And where the consensus in the scientific community is as strong as it was in this case, then it is hardly the news media that is the reason for the public wondering whether or not stricter controls are needed."

Cline replied: "I guess I would make one comment, Congressman, and that is, advances in science are generally not made by consensus."

The hoopla over human gene transplants took some of the thunder away from another first in the genetic engineering of live animals announced in the fall of 1980: the development of genetically altered mice from embryos that had been injected with foreign DNA.

18. Injecting Embryos

Each new power won by man is a power over man as well.
—C. S. LEWIS, 1965

The operation looks almost gentle in microscale. A hollow glass needle nudges against the outer membrane of a tiny egg cell, transparent as a soap bubble. The membrane bends, resists a second, then gives way as the needle stabs into the gelatinous cytoplasm. One more barrier to pierce, the nuclear membrane, then a minute tap on a syringe propels a surge of liquid carrying thousands of genes. The nucleus swells and the needle quickly withdraws.

The procedure is more brutal than it appears. One scientist who has spent hours at the microscope guiding the needle compares it to "your being speared by a telephone pole." It is a massive insult. Yet cells are not soap bubbles. Unless a researcher gets too heavy-handed, the majority of eggs will be no worse for the invasion. A few will be changed for life, and a very few will pass that change on to generations of descendants.

Since 1980 this technique, called "microinjection," has given scientists their first undisputed successes at transplanting genes into living animals. Of the three major methods available today for transferring genes, microinjection is the most "unnatural," a procedure microbes apparently never thought of or had the equipment to carry out. (Although a virus attaching its hollow syringelike tail to the surface of a bacterium and sending in its DNA to take over the cell comes a close second.) Microinjection is a human invention, direct, unsubtle, and surprisingly effective.

The roots of the technique go back at least thirty years to the beginnings of work most laymen term "cloning." The scientists doing it called it "nuclear transplantation." What they were doing with needles then was pulling the entire nucleus with its full complement of genes out of an egg and substituting a new nucleus. Some of the questions they were trying to answer remain basic still: Does every cell in a creature retain the same full set of genetic instructions that came in the fertilized egg? At what stage in development do cells begin to "differentiate," choosing a specialty such as a kidney cell, brain cell, eyeball, or fingernail, and shutting off all genes but the ones needed for that role? When is the commitment to a specialty final? Can it be reversed? Can the turned-off genes of a specialized adult cell be turned back on again?

A British biologist, John B. Gurdon, dramatically confirmed the answer to the first question in 1962 when he pulled the nucleus from the egg of a South African frog and replaced it with one from a cell in the gut lining of a tadpole. A tadpole obviously is not a mature adult frog, but the cells in its gut lining had already closed out some options for what they would do when they grew up. Cell specialization had begun. But the result of the experiment showed the process was not too advanced to be reversed. The gut-cell nucleus gave up its specialty plans and directed the growth of the egg into a normal adult frog.

Like the nuclei of all cells, the nucleus of the gut cell carried a complete set of genetic instructions for building its kind of creature. The instructions directed the egg to divide normally, generating daughter cells that differentiated into all the specialized cells needed by an adult frog.

Scientists around the world are continuing to replace the nuclei of eggs with nuclei from older embryos, fetuses, or adult tissues in an effort to determine exactly when and how the full potential of the genes in a specialized cell gets shut down.

A clone, as we know, is a genetically identical copy. Cloning molecules like genes has become routine with the genetic engineering techniques developed in the past ten years. Other types of clones are common, too. A geranium grown from a cutting is a clone of the mother plant. The frogs that grew from Gurdon's engineered eggs were clones of the tadpole that donated the nuclei. A man has not yet been cloned, despite books and claims to the contrary. It was 1981 before a team of scientists claimed to have achieved the first successful cloning, of a sort, in mammals.

Karl Illmensee of the University of Geneva, Switzerland, and Peter Hoppe of Jackson Laboratory in Bar Harbor, Maine, reported they had produced mouse clones. The procedure was not substantially different from what Gurdon had used in frogs, except in scale and delicacy. A frog egg is large enough to be seen with the naked eye, about as wide across as the tip of a sharpened pencil. A mouse (or human) egg cell is 100,000 times smaller. The team reportedly took gray mouse embryos that had grown to the thirty-two- or sixty-four-cell stage, separated the cells, and used a glass pipette with a microscopic tip to pull the nuclei from a number of them. Then they injected these gray nuclei into fertilized black mouse eggs and pulled out the eggs' own nuclei. These black mouse eggs loaded with gray mouse nuclei were implanted in white surrogate mother mice. The reported result: Several gray mice were born, identical to the embryos that had donated the nuclei.

In 1983, however, Illmensee's claims were called into question by a number of his colleagues. While the charges were being investigated, another team led by James McGrath and Davor Solter at the Wistar Institute in Philadelphia reported it had achieved the same thing—the production of mice by nuclear transplants.

Two or more individuals developed from nuclei taken from the same embryo are clones of each other. The only catch to this type of cloning, for humans enamored with the possibility of perpetuating identical copies of themselves, is that the original embryo is destroyed in the process of donating nuclei from its still meager complement of cells. Few of us may want to be sacrificed at the fourth day of embryonic life in order to be resurrected in triplicate. So far no one has succeeded in cloning a mammal using nuclei taken from an adult, a stage when the donor can easily afford to give up a few skin or blood cells.

It quickly became apparent that microinjection techniques could be used for moving other things than nuclei in and out of cells. In 1970 Elaine Diacumakos at Rockefeller University modified the techniques and apparatus and began injecting bulk DNA into cells. Others injected copies of genetic instructions—messenger RNA molecules—from various species into frog eggs, learning something about the universality of the genetic code in the process when they found that frog cell machinery knows how to handle genetic instructions to produce honeybee venom or rabbit hemoglobin.

In the late 1970s, when molecular biologists began to move purified genes into cells by using viral carriers or DNA transformation, it wasn't long before some decided to try microinjecting naked genes directly into a nucleus.

In 1979 and 1980 at least three groups successfully injected viral genes into mouse cells and ended up with cells that produced a viral enzyme. The success rate was much higher than with DNA transformation. Microinjection had the advantage of getting the genes directly into the nucleus, past the barrier of the cell membrane and the enzymes waiting to destroy them in the cytoplasm. It was an efficient but tedious way to move genes into individual cells. But there was one group of cells for which the return might be worth the tedium: the fertilized egg, the one stage in a creature's life when successfully injecting new genes into a single cell could create an adult with foreign genes in every cell.

The prospect of producing genetically altered strains of mice and other animals held strong appeal, and not for Frankensteinian motives. Whole colonies of animals carrying certain mutant genes could serve as models for human disease, allowing scientists to study defects and try out new drugs and therapies. A foreign gene inserted in an animal's chromosome could serve as a flag by which to follow the fate of various traits as tissues developed and became specialized. And everyone was aware of the possibilities the technique could provide for introducing new traits and capabilities into livestock.

No other gene transfer method seemed to offer these possibilities. In DNA transformation the low percentage of cells that picked up new genes—one in 10,000 at best, often one in a million—made this method too wasteful to use on embryos. It would take thousands of pregnant mice to yield enough embryos to get one successful transformation. And the experiments that Martin Cline tried—removing a limited number of cells from an adult mouse or human, transforming them with new genes, then infusing them—could not be used to get new genes into every cell. In particular, this method could not get new genes into the germ cells—the eggs and sperm—so that they could be passed on to future generations of animals as a permanent part of the lineage.

Microinjection offered this promise, and so in the late 1970s work began in earnest to inject new genes into embryos.

No one had done this with isolated genes before, but there was reason to believe it could work. Since 1974 Rudolf Jaenisch, now of the University of Hamburg, Germany, and his colleagues had been infecting early-stage mouse embryos with viruses—in effect, strings of foreign genetic material—and reimplanting the embryos in foster mothers. Since the embryos had already divided beyond the single-cell stage by the time of infection, not all the cells received viruses. When the animals were born, many were mosaics. They carried viruses in some but not all their tissues. Many carried the viruses in their germ cells, however, and passed on these bits of foreign DNA to their offspring. Because these second-generation animals grew from fertilized eggs containing the virus, they carried the foreign material in every cell of their bodies.

In 1980 Jon Gordon and Frank Ruddle at Yale University injected cloned viral genes directly into the nuclei of fertilized mouse eggs. The gene used—for the enzyme TK—provided no useful trait that they wanted to confer on an animal. It served simply as a "marker card" in a stacked deck—something they could follow easily to see if they were achieving their goal. They were. Of the 180 mice born from injected embryos, three carried the viral genes in their cells. The genes were not intact, and they did not seem to be functioning, but they were there.

The announcement came shortly before the news of Cline's human gene therapy attempts, and it marked the first undisputedly successful transfer of specific genes into living animals.

In 1981 at least four groups reported they had taken the process one step farther, transferring mammalian genes into mice and getting them passed on to second and third generations of animals. But it proved more difficult to get the genes to work properly in their new settings. Ruddle and Gordon injected the human gene for interferon, an antiviral protein, into mice and watched it passed along to a third generation of animals that year. The gene did not function, however.

A team at England's Oxford University, Frank Constantini and Elizabeth Lacy, inserted a gene for rabbit hemoglobin into mice and found that some mice passed on the gene to their progeny. None of the mice seemed to use the transplanted genes to make the rabbit blood product in their blood-forming cells, although tests later showed that two animals had apparently activated the genes in an inappropriate place. The animals were trying to make hemoglobin components in their muscle cells.

A flurry of headlines greeted the announcement by another research team (headed by Thomas Wagner of Ohio University and Peter Hoppe) that they had also successfully injected rabbit hemoglobin genes into mice and observed them carried into a second generation. The difference was that the team also claimed evidence that some of their animals in each generation were making at least low levels of rabbit protein in the right places, the blood-forming cells, and not in any other tissues. It sounded like the breakthrough for which researchers in gene transfer had been waiting. Wagner, who had ties to a commercial firm called Genetic Engineering, Inc., of Denver, Colorado, told reporters the company was already at work on "three-parent cattle"—cattle with foreign genes added to those from the natural parents.

But the team's claims of a transplanted gene that worked met the same fate among molecular biologists that Martin Cline's animal experiments had the year before. Scientists examining the telltale blots and bands from the chemical analyses were not convinced the transplanted genes were working.

Only one group in 1981 seemed to have convincing evidence that a foreign gene, injected naked and unassisted into embryos, was active, and the gene was from a virus, not a mammal. A team led by Beatrice Mintz of the Institute for Cancer Research in Philadelphia had injected mouse embryos with both human hemoglobin genes and viral TK genes. Five mouse fetuses they later examined carried intact copies of both genes. There was no evidence that any of the hemoglobin genes were active, but one of the five animals had put the TK gene to work and was making significant amounts of the viral enzyme. Not only had alien genes been transferred successfully into a living animal, they had actually started to function there. The results fueled the hope that the genetic endowment of animal lineages could be altered by human manipulation. In another lab, results with injections of hybrid genes were about to make that hope a reality.

All the research teams working with microinjection use essentially the same procedure, and yet the number of eggs that tolerate injection, survive to birth, carry the foreign genes, and put them to work varies greatly. The real mystery seems to be why the procedure ever works at all, and scientists still haven't sorted out the factors that will enhance its success or efficiency.

The process begins by letting a female mouse become pregnant in the time-honored way. A few hours after mating, the female is killed and the fertilized egg removed to a glass lab dish. At that point the sperm has entered the egg, but it has not fused yet with the egg nucleus. These two objects are called "pronuclei" at this stage. It takes about eight hours after mating for the two pronuclei to fuse, combining a random selection of half the genes from each parent into a unique full set of genes and triggering the development of a new individual. This is a special, vulnerable moment in the life of an organism, the only time when it is programmed to accept an infusion of foreign DNA.

With a hollow glass needle, scientists go for one of the pronuclei. Both look the same, but the male is closer to the surface of the egg and most accessible, so it will very likely be the one to be injected. Scientists haven't devised any technical tricks to ensure that once genes are injected, they will get processed and integrated into the chromosomes. For this crucial step they must rely on the egg, hoping the extra DNA will be caught up in the bustle of the genetic merger, welcomed, and processed as though it came from one of the parents.

"We don't know how gene transfer works, but it's probable that you need to saturate the mouse nucleus with the gene so that it can find the enzymes that are needed to get it into the chromosome," says Jon Gordon, now at Mount Sinai School of Medicine in New York. "We've put as many as thirty thousand copies of a gene in a mouse embryo. That's like adding a chromosome. But other people have gotten very high rates of transfer with only five hundred copies."

The number of eggs that survive the procedure depends largely on the skill of the researcher. The eggs are so small that a hundred of them would be barely discernible to the eye. "You can kill anywhere from five to seventy-five percent on any given day," says Gordon. "On a good day I'll only kill twenty-five percent." A little too much pressure on the heavily downgeared instruments and the researcher can stab all the way through and out the other side of a nucleus, or can squirt in so much gene-bearing liquid that the nucleus bursts. The glass needle itself—as small as a micron across at the tip—may tear the delicate nuclear membrane as it is withdrawn, pulling out strings of DNA in a sort of genetic disembowelment. The eggs that survive the injection are implanted through an incision into the oviduct of a surrogate mother. The number that survive to birth is variable. Even among natural embryos, only 50 percent survive.

By the end of 1982, major researchers had become proficient enough that 20 to 40 percent of the animals born from injected eggs not only carried foreign genes but could pass them on to their offspring. Still, in most experiments the genes worked fitfully if at all. Beatrice Mintz's group had two lines of microinjected mice—one carried human hemoglobin and viral TK genes, and the other carried human genes for growth hormone. The genes were heritable, but they didn't function.

Constantini, now at Columbia University, and Lacy tried injecting a hybrid gene made by splicing together pieces of mouse and human genetic material—the control sequences from a mouse hemoglobin gene and the protein-coding section of a human hemoglobin gene. Maybe the mouse cells would recognize the mouse control switches and use them to turn on the human gene. They didn't. The animals did, however, keep the gene and pass it along to their descendants.

But the idea of a hybrid gene paid off spectacularly in 1982 for another group of researchers. The team from four institutions working together succeeded for the first time in getting a foreign gene turned on full blast in an animal. The results were dramatic—mice that grew to twice their normal size—and the scientists proclaimed the opening of "a new era in genetic engineering."

19. Building Mighty Mice

Since we surely now know that scientific research, whether
basic or applied, is a source of enormous power for both
good and ill, the scientific researcher has, then, an
obligation to be as active in his moral imagination as in his
scientific imagination. We ask the same of any person in a
position of power.

— DANIEL CALLAHAN, 1976

An animal responds to stress and physical trauma on many levels,
from mental to molecular. One of the most basic responses is the
switching on of a gene that makes a protein called metallothio-
nein, or MT. Chase a mouse around its cage for five minutes, dose
the animal with high levels of toxic metals, or infect it with bacteria, and it will switch on the MT gene in tissues throughout its
body.

When the MT gene is isolated and transferred into cells in a lab
dish, it loses its ability to respond to some of these insults. But the
gene can still be switched on by exposure to heavy metals such as
zinc and cadmium, producing a protein that binds with the metals
and helps the cell to sequester and tolerate the excess.

Researchers were intrigued. Here was a gene with a powerful
control switch that they could easily trigger at will. Perhaps the
switch could be removed, linked up to other genes, and used to
turn them on in living animals with the same power and efficiency
that viral signals have demonstrated in forcing hybrid genes to
work in cultured cells.

Ralph Brinster of the University of Pennsylvania and Richard
Palmiter of the University of Washington, Seattle, began testing
this idea by hooking the control signals from a mouse MT gene to
that old standby in gene transfer work, the viral gene for the en-

zyme TK. They injected this hybrid gene into mouse embryos. At the time the two researchers began, only one mouse of all the dozens raised by various researchers from embryos injected with "naked" TK genes had ever been found to produce the viral enzyme. The results with the hybrid gene were more dramatic: Three fourths of the mice that carried the foreign gene produced low levels of the viral enzyme without any coaxing.

When Brinster and Palmiter treated the animals with nontoxic doses of cadmium or zinc, the mouse MT gene controls turned up the production levels of the attached viral genes. With the MT control signals in charge, the TK genes acted as though they were MT genes, producing the highest levels of enzyme in the liver and kidney and much lower levels in other tissues such as the brain. This production pattern is the same as that seen for MT.

Scientifically the event was significant. Transplanted genes were working vigorously in a living animal for the first time; and were inherited and remained active in a number of the offspring. But TK is a rather mundane if necessary enzyme, not a substance that could have a profound or visible impact on the animals. The next hybrid gene the team injected was different. When it was active, the effect was unmistakable and the results made headlines.

Working with colleagues Michael Rosenfeld at the University of California at San Diego and Ronald Evans at the nearby Salk Institute for Biological Studies, Brinster and Palmiter hooked the mouse MT controls onto a rat growth-hormone gene. Hundreds of copies of the spliced gene were injected into each of 170 mouse eggs, which then were implanted into surrogate mothers. Twenty-one mice were born, seven of them carrying copies of the foreign gene.

Six of the seven began to outgrow their littermates. After the mice were weaned, researchers began to supplement their diets with zinc to activate their MT control switches. By the time the animals were three months old, some of the fastest growing ones were twice the size of the mice without hybrid genes. When these were bred, they produced a second generation of giant mice carrying the hybrid gene throughout their bodies.

The mice grew bigger not because rats are bigger than mice or the rat growth hormone is more potent, but because the gene with the MT controls attached was active in many tissues and couldn't be switched off by the signals the mouse uses to control its own growth hormone genes. Normally growth hormone is produced only in the pituitary gland at the base of the brain, and output is regulated by other hormones. But in these mice, the bulk of the growth hormone was made in the liver. The liver is where the MT gene is generally most active. The mice that had retained the largest number of injected genes—ten to thirty-five copies in every cell—produced the greatest blood levels of the hormone and grew to be the largest. Their blood had about 800 times the normal level of growth hormone.

When a defective gene causes that kind of runaway growth hormone production in the human pituitary gland, the result is a disease called giantism. (The opposite defect in the gene leads to a type of dwarfism.) The research team immediately proposed that their oversized mice could serve as models to study that defect.

But the work has more far-reaching implications, and the researchers were fully aware of them. One prospect would be to use a hyperactive growth hormone gene to produce rapidly growing strains of cattle, pigs, sheep, and other livestock which could reach market size in half the normal time.

Another promising possibility is splicing the MT controls to human genes and transplanting them into cattle for "gene farming." In principle this would be similar to commercial genetic engineering ventures already using bacteria to produce human insulin, interferon, and other pharmaceuticals. But it would open the way for the mass production of human proteins that bacteria are not equipped to process. Many proteins such as blood-clotting factors that could be useful in treating inherited diseases must be modified chemically in special ways by a cell after they are produced before they can become active. Microbes do not have the genes or the machinery to perform some of this work, but the cells of cattle and other mammals probably do.

Palmiter envisions injecting human genes for useful proteins into cattle embryos and creating new strains of animals that would pass the human genes on to all their progeny. "Genes for products that are normally secreted into the blood are the easiest to imagine," he says. "You could bleed your cow every three weeks just like people used to do to harvest antiserum [antibodies] from goats and horses, then purify the protein. So you're basically farming gene products. I imagine it'd be cheaper to put a cow out to pasture than to grow umpteen liters of bacteria in culture. And it might be the only way to make certain products."

Palmiter doesn't make those predictions glibly. The task will not be technically easy in cows even though it has already been done in mice. The low efficiency of the current embryo injection process is a major problem, and scientists have no idea how to enhance their success rate. Only 10 percent or less of injected mouse eggs survive to birth, and fewer than half of those mice can be expected to carry the transplanted gene, put it to work, and pass it on to their offspring. And the genes do not always remain stable and active from one generation to the next.

Just collecting large number of eggs from cattle is more costly and difficult than from mice. Besides, Palmiter notes, a researcher microinjecting transparent mouse eggs can guide the needle into the nucleus and see where he is putting the new genes. Cow's eggs are opaque.

"I think there may be years of hard work trying to figure out how to make the technique superefficient in the mouse and hoping that it will then apply to other animals of interest at some future time," he says.

The team, like most researchers in the field, quickly downplayed any talk of microinjecting hybrid genes into human embryos to correct inherited defects. Transplanting genes into selected tissues of a patient is morally comparable to implanting a new kidney, heart, or liver in his body. But injecting genes into a fertilized egg—genes that with any luck will be carried in every cell of the individual—raises unique ethical and philosophical questions about the wisdom of making inheritable changes in the human gene pool. (Although, as geneticist Bernard Davis of Harvard University has pointed out, permanently correcting a genetic defect would have the same effect on the human gene pool as prenatal screening and abortion of defective fetuses. In either case, the frequency of the defect in future generations is reduced.) Aside from these considerations, the logistical problems are still overwhelming.

Removing human eggs from the womb and reimplanting them present no major obstacle. More than a dozen test-tube baby clinics around the world already do this routinely. But at best, one in twenty eggs microinjected and reimplanted today would produce a live baby with active foreign genes. Human embryos are certainly less expendable for such an inefficient experiment than cow's eggs, and few couples are likely to go for those odds. The techniques, however, will eventually be improved.

Another consideration, and one less easily overcome, is that now it is impossible to detect genetic defects in a single-celled embryo. Current screening tests require the collection of a number of cast-off fetal cells that can be grown in a lab dish until the mass of genes, chromosomes, and protein products is substantial enough to be examined.

If a way were somehow devised to spot a defective gene in a single fertilized egg cell without destroying it, a new question would arise: Why bother to fix and reimplant a defective egg in the mother's womb? It would be far more practical to discard the defective egg and test another one. Even if a parent suffers from a dominant genetic disease like Huntington's chorea, half the embryos should be normal. For couples with a recessive disease, each embryo has a 75 percent chance of being normal. More embryos would actually make it alive through this kind of screening process than would survive being "fixed" with current microinjection techniques.

Injecting genes into a later, multicelled embryo or fetus would produce a mosaic, a person with the foreign gene in only some of his cells. This could theoretically provide enough active genes in random tissues to help correct some inherited defects. But there are very few gene products that humans need to have pumped into their tissues at random with the same uncontrolled exuberance the MT signals display in churning out growth hormone.

"Let's say you're missing any one of a number of crucial serum proteins and the gene for it was available," Palmiter says. "If it didn't matter what cell in the body made that protein for you or whether it was regulated, I guess you could conceive of technically doing it with the tools that are currently available.

"But most things you'd want to be regulated properly. Too much is as harmful as too little. If you wanted expression of growth hormone only in the pituitary and only in response to the proper hormonal stimulation, that would be difficult to achieve with what we know today."

Research over the past five years has made it dramatically clear that we have several effective options for transplanting genes into single cells or living animals. Some transplanted genes seem to function occasionally and fitfully on their own in new cells. Others can be tricked or forced into working with varying degrees of enthusiasm when they are hooked to strong control signals from viruses or animal genes. The challenge of the 1980s is learning how to regulate transplanted genes so that they perform normally, a step that will provide us with a realistic chance of correcting genetic defects in humans.

20. The Genetic Bureaucracy

> What distinguishes a butterfly from a lion, a hen from a fly,
> or a worm from a whale is much less a difference in
> chemical constituents than in the organization and
> distribution of these constituents . . . What makes a
> chicken wing and a human arm different is not so much the
> material out of which both are made, as the instructions
> specifying the way one or the other is built."
> —FRANÇOIS JACOB, French biochemist, 1982

DNA, unloosed from the confines of a cell, is deceptively uncom-
plicated. To the naked eye it is a tiny mass of clear, sticky threads
that can be wound like cotton candy onto the tip of a glass rod, an
identical yard of microscopic threads in the nucleus of each
of our one hundred trillion cells. Metaphors for DNA are one-
dimensional: a computer tape or a string of beads, all the genetic
information that shapes and defines us laid out neatly in a linear
array.

Viewed that way, it seems but a simple problem to add a few
more genetic beads to the string—new genes sent into the nucleus
on a virus, deposited there with a needle, or slipped through the
cell wall and shepherded across the cytoplasmic frontier while the
enzymatic guard is down; new genes that will change the life of
the creature that receives them. But it's not that simple.

DNA alone is not life. DNA is the instructions, the code, the
clichéd "master blueprint" of life. Life is that code implemented
in space and time. Life requires bureaucracy, architecture, sched-
ules, assembly lines, a constant flow of energy, materials, and
information. Variations in scheduling can make an enormous difference in the sort of creature that gets built and operated. Only a

1 percent difference in DNA separates us from apes; the striking differences are largely a consequence of changes in a few genes that control the timing of our development. We have been called "apes that never grew up," products of an extended childhood, a delayed physical maturity that left us with more time for brain growth, socialization, and learning.

Life's smallest functional unit is not the gene but the cell. A cell is often compared to a walled city, its integrity guarded and traffic in and out of its gelatinous preserve limited by its sentry-covered cell membrane. Within the cell's cytoplasm bustle the industry and commerce of life—mitochondrial power plants generating energy, ribosomal factories manufacturing proteins, digestive sacs called lysosomes breaking down food molecules and invading microbes.

In the middle of this bustle stands the government center, the nucleus, and within its membrane are the bureaucrats that run the place—the genes, spiraling double strands of DNA coiled and recoiled inside their cramped quarters like compressed springs. Just before cells divide, the DNA is bundled into the pairs of thick, rodlike structures we know as chromosomes. But most of the time, DNA in the nucleus exists in another form. It is spooled around tiny beads of protein. The beads and the strands of DNA between them are spiraled into larger coils, and these in turn are looped into supercoils. The protein beads and DNA together are called chromatin.

In any given cell, the vast majority of bureaucrats do nothing. Each cell of a higher animal has its assigned tasks to perform for the benefit of the larger community, and only the small corps of genes needed to keep up the housekeeping functions of the cell or run its specialized tasks are active.

If a foreign gene is to have an impact on how the place is run, it must learn somehow to fit into the active bureaucracy of a specific cell. And the infiltration must succeed not just once, but in a significant percentage of cells if it is to have an impact on the life of the whole animal—the goal of genetic engineering.

Finding out what it takes to be an effective working member of the genetic bureaucracy in a particular type of cell is one of the central goals of modern molecular biology. Future gene therapy depends on being able not only to get new genes into cells, but also to get them regulated and functioning as active members of this team—either by mimicking the natural system or by technical trickery.

* * *

Tens of thousands of proteins, each coded for by a gene, carry out the functions of a living system. A gene cannot act on or even deliver the instructions it embodies. To express itself and have a proper impact on the system, a gene must simply be in the right place at the right time, and in a properly receptive mode so that the swarms and ranks of chemical functionaries in the cell can "read" it and see that its instructions are translated into protein. In the thirty years since the spiraling ladder structure of DNA was discovered, scientists have outlined the major events involved in this flow of information within the cell. The more they have learned about the process, the more complex it has turned out to be.

First, a working copy of the gene is made for delivery to the factory. This process is called "transcription," and it requires the twisted double ribbon of DNA to unzip at the site of the gene. The genetic sequence on one of the strands is transcribed or copied not into DNA but into the related four-letter code of RNA. Then the copied message, the messenger RNA or mRNA, has to be edited. This is a fairly recent revelation. Not until the mid-1970s did molecular biologists realize that most of the genes of higher creatures, unlike those of bacteria, are not continuous sequences of DNA. The DNA sentence, as we saw in Chapter 10, is split into phrases—the protein-coding sequences—interrupted by one to fifty or more stretches of unintelligible garble called intervening sequences, or introns. These must be censored—chopped out of the mRNA—and the code-bearing phrases strung together into a coherent sentence.

Once the message is processed and spliced, and a chemical "cap" and "tail" added front and back, it is transported out of the nucleus to the protein factories, the ribosomes. There the message must be translated into another alphabet—the twenty-symbol code for the twenty amino acids from which proteins are assembled. The ribosomes, structures measuring about one millionth of an inch across, attach to the mRNA strand and begin moving down it like assemblyline workers, reading the code, grabbing the specified components, and assembling an ever-lengthening protein chain anywhere from 50 to 1,500 amino acids long. The reading or the decoding and grabbing are actually done by another of the many chemical intermediaries in the cell, transfer RNA (tRNA). Special triplets or groups of three molecules in the mRNA code tell the tRNA molecules where to start reading the message and where to stop and let the ribosome release the finished protein.

Reduced to the singular, the process looks complex enough, but in reality it never occurs in the singular. Genes do not work one at a time. Of the fifty thousand to one hundred thousand genes in a human cell, perhaps ten thousand remain active in the average specialized cell, switching on and off as needed to let muscles flex, red blood cells bind oxygen, or nerves transmit impulses. On a submicroscopic scale, the frantic choreography of any split second must rival rush-hour traffic in Manhattan.

The tightly packed and coiled strands of DNA wind and unwind at thousands of sites, unzipping and zipping back together as they are copied. Legions of mRNA copies are edited, spliced, and hustled off into the cytoplasm. More than 100,000 ribosomes run down the linear messages, stringing dozens of links per second onto protein chains. The proteins are shuttled off to run the hundreds of simultaneous biochemical cycles and cascades that fuel life. Among the proteins are legions of enzymes and other intermediaries bustling everywhere, getting each operation stopped and started, checked and rechecked, signaling, scheduling, and coordinating endlessly.

The successful expression of a single gene's message thus involves numerous control points where the rate, timing, accuracy, and efficiency of the process can be regulated. Little wonder that most foreign genes, sent in naked and unprepared, get lost in the shuffle and are never heard from again! At first scientists concentrated on just proving that the genes had arrived in their new settings. But as that step—gene transfer—became routine, they began to concentrate on how, where, and with what control signals foreign genes should arrive in order to have a shot at becoming fully naturalized citizens.

Molecular biologists began their search for the secrets of gene regulation in microbes, and by the end of the 1960s, they felt they had a pretty good idea of how things work in bacteria. The principle control point in microbes turned out to be at transcription, the beginning stage at which working copies of the DNA blueprint are made and sent off to be translated into protein. A simple on-off switch called a "promoter" runs the show, starting or stopping the copying process depending on signals from "inducers" or "repressors."

The bacteria *E. coli*, as we have seen, produce an enzyme (beta-galactosidase) that helps them break down and draw energy from lactose, or milk sugar. The gene for the enzyme is switched on and working only when it is needed—when the sugar is present. It is switched off by a repressor chemical that clings to it and prevents it from being copied. A constant supply of the repressor is produced by another nearby gene. But the bacteria obviously do not want the gene repressed all the time or they could never "eat" milk sugar. How do the bacteria distract the ever-present repressor when sugar is available and they need a supply of beta-galactosidase?

The sugar itself does the job. The sugar is attracted to the repressor and pulls it away from the gene. The gene begins churning out orders for its enzyme, and the enzyme starts breaking down the sugar. As the sugar is used up, the repressor is freed, settles back onto the gene, and shuts down the cycle.

Bacterial control switches hold more than academic interest for scientists. They are central to the goal of converting bacteria into pharmaceutical factories by "teaching" them to produce human hormones and enzymes. By splicing bacterial control switches onto human genes and inserting these hybrids into bacteria, researchers have successfully tricked microbes into using their own cellular machinery to produce insulin and other human proteins.

The next question molecular biologists had to tackle was whether gene regulation in higher animals follows the same rules as in bacteria. As it turns out, the dominant control level for animal genes does seem to be at the same stage, the beginning of the transcription process where mRNA copies of the gene are made. But animal genes have other complex factors to contend with in getting themselves expressed that one-celled creatures without nuclei do not. First a processing-and-transport system must chop out interruptions in the mRNA, splice the message back together, and slip it out through a pore in the nuclear membrane before translation and protein manufacturing can begin.

A large genetic bureaucracy also has historical and political realities that affect how each individual gene will perform. Events that occurred during the development of the organism have given each cell a specialized assignment and caused it to activate a select group of genes. Family ties have evolved between related genes. And the packaging of genes into neighborhoods and architectural structures within the nucleus has affected their roles.

The easiest of all these control points to tackle was obviously the promoter, the on-off switch. If it followed the pattern of bacterial genes, the signal had to be somewhere in the DNA sequence itself, and that was something scientists had learned by the early 1970s to manipulate. They began knocking out bases, the chemical units of DNA, one or more at a time from the front end of animal genes to see which were necessary for the gene to continue being expressed.

Animal promoter signals have not turned out to be as clear-cut and as close by the genes as bacterial promoters, and scientists still have not firmly defined them all. For instance, thirty chemical units "upstream" from most genes is a site researchers call the "TATA box" or the "Goldberg-Hogness box." This site seems to provide a flag to tell enzymes precisely where to begin copying a gene into mRNA. It's clear that upstream from the TATA box there are other signals that seem to be important in gene regulation and function. These differ from gene to gene. Downsteam from some genes, scientists have identified enhancer elements that seem to increase the rate at which mRNA copies of the gene are made.

Even without a full understanding of promoter sites, scientists in the late 1970s and early 1980s have been able to exploit this control region to get foreign genes expressed in both cells and living animals. Viral promoters spliced onto bacterial or animal genes allow almost any gene to be turned on at some level in animal cells. And the use of a promoter region from a mouse gene spliced onto a rat gene, as we saw earlier, unleashed uncontrolled expression of the gene and resulted in the production of giant mice.

The nature and relative importance of other gene control factors have been even harder to pin down than promoter sites. The linear nature of DNA sequences makes it tempting to think of genes as one-dimensional objects. But genes exist in 3-D, and structure and packaging apparently play a role in their function. Some research suggests the coilings of the chromatin may bring control signals like the TATA box and even more remote control sites into some sort of functional alignment. Researchers have also found that the chromatin structure seems to be more relaxed and open in neighborhoods where genes are active, an arrangement that could leave the genes exposed and accessible to copying enzymes. In contrast, inactive genes tend to stay tightly bundled, hidden in proteins.

But what determines which genes will be active and ready to

function in any particular cell? The answer may involve some kind of chemical commitment made by a cell during the animal's development. Something in the developing organism "knows" that an eyeball won't need to make insulin and nerves won't make sex hormones, and it begins to shut down categories of genes as tissues differentiate. In more primitive stem cells, such as those that divide to produce renewed supplies of a variety of blood cells throughout life, larger groups of genes may be left in readiness until specific commitments are made.

How are genes shut off? Experiments show that some inactive genes have extra chemical baggage called methyl groups slapped onto their bases. Some foreign genes injected into early embryos get methylated and never become active, although others such as the hybrid gene that produced the giant mice do not receive this treatment. No actual cause-and-effect relationship between methylation and gene activity has been shown, and the question remains largely unanswered.

At another level of control, families of related genes must be coordinated so that red blood cells have enough alpha-globin chains to match with beta-globin chains to make hemoglobin molecules, and the body's defense system can assemble antibody molecules tailored to latch onto specific foreign invaders. How do genes "talk" to each other? One theory is that the discarded bits of mRNA chopped out of the genetic message after it has been copied serve as a signal to other genes that this gene is at work.

How important is each of these control points? The answers are incomplete, but they are coming. Understanding human gene regulation is the primary motivation behind most gene transfer experiments today. Researchers are inserting genes into places where they would not normally be to see what it takes to make them work there.

Not all genes are difficult to turn on when they are transferred into foreign cells. One group called "constitutive" genes seems to be perpetually on at constant levels, even without the help of powerful controls spliced onto them. Genes for housekeeping enzymes like TK that are produced in every cell, for example, can be moved from viruses, bacteria, or animal cells into those of other species and resume their work as though they were right at home. Other genes are normally expressed only when they are exposed to certain external stimuli, and they will often respond at some level to this challenge even when put in a foreign setting. Heat-shock genes from a fly transferred into mouse cells, for instance, will still jump into action when a critical temperature is reached.

But the most interesting genes to both basic scientists and medical researchers are those such as the hemoglobin family that require complex regulation. The genes work only in specialized cells; various members of the family switch on and off during different stages of embryonic, fetal, and adult life; and each gene must coordinate its activities with those of related genes. These are the sorts of genes that must be fastidiously regulated, allowed to produce their products only when and where and in the amounts needed or the results can be dangerous or monstrous. Giant cattle may prove useful to man, but giantism is a disease when growth hormone genes run amok in a human.

In the past few years, scientists have sometimes coaxed such specialized genes to work in new settings unaided by artificial controls, but not yet in a normal way. The results are encouraging, but molecular biologists would probably not want to turn these genes loose in their own or any other person's cells yet. Even putting the right kinds of genes that have some of their natural control signals attached into the right kind of environment has proved to be not enough to initiate normal gene activity.

One of the types of laboratory-grown cells available to molecular biologists is a line of mouse cells that are destined to become red blood cells. The cells—murine erythroleukemia, or MEL, cells—have had their development blocked by viral infection. They haven't yet begun to perform the duties of a red blood cell, specifically the production of hemoglobin. A number of chemical inducers like DMSO will unblock the MEL cells and let them mature and perform like red blood cells. Scientists do not know how or why this works, but it is convenient.

Richard Axel of Columbia and Thomas Maniatis of Harvard took MEL cells and incubated them with human beta-globin genes, the same genetic clone used by Martin Cline in his 1980 attempt to implant new genes into two thalassemia victims. Some of the mouse cells picked up the foreign genes. When the researchers doused these transformed cells with DMSO, three fourths began to put not only their own but the human genes to work turning out instructions for hemoglobin.

But there is a catch. The mouse cells' own genes produced a much larger quantity of mRNA copies than did the foreign genes.

The researchers had left attached to the human beta-globin genes an extra segment of DNA that includes its natural promoter region. But this control sequence was not enough to get the genes to behave like the mouse cells' own. The same thing happened when a hybrid gene was constructed, with promoter regions from a mouse globin gene spliced onto the human globin gene. The hybrid performed, but not at an equal level with the mouse cell genes. The natural globin promoter was not enough to get the gene to work normally, and neither was being put into an environment where it was supposed to function—a red blood cell.

"Something is still different and we don't know what it is," Axel says. "I think it has to do with an event in the developmental history that the exogenous [foreign] gene missed out on, and we don't know what that is." Or, Maniatis speculates, subtle but critical promoter signals even farther up or downstream from the gene may have been left out when the gene with its nearby promoter region was isolated and cloned.

In other experiments with very different results, foreign genes also seemed to be missing out on some regulatory experience that had affected the cells' own genes. In this case, the foreign genes worked in places where the normal genes do not. When Axel transferred a human growth hormone gene with its adjoining control switches into mouse skin cells and exposed the cells to steroid hormones, the human gene went to work just as it would in the pituitary gland. The cells' own growth hormone genes, inactivated earlier in development when the cells chose to specialize in skin rather than in pituitary activities, remained dormant. (The hybrid growth hormone gene used to create giant mice also worked in abnormal tissues, but for a different reason: It was under the control of a promoter from the MT gene and therefore was activated in tissues where MT is normally produced.)

"While qualitatively these genes put into cells seem to respond to normal control signals, there's still something different about them," Axel says. "In the globin case, the difference was quantitative. They made much less. In the growth hormone case, it responded to the right regulatory signals, but it was inappropriately expressed. In other words, it shouldn't have responded or been made in that cell at all. It thought it was in the pituitary."

In 1983 Japanese scientists announced a partial success in getting genes powered by their own natural control regions to work most efficiently, if not exclusively, in the correct specialized cells. A team at Kyoto University transferred a gene for a protein produced in the lens of a chick's eye into mouse skin, brain, kidney, liver, and lung cells. Few, if any, of these cells made the protein. But when the genes were put into cells taken from the lens of a mouse's eye, 40 percent of the cells made the chick-lens protein. When the team replaced the gene's natural control region with signals from a virus, the gene was induced to work in all cells, even the inappropriate ones, as expected.

One hope for overcoming the remaining hurdles to normal gene regulation is to find a way of targeting genes to precise locations on the host cells' chromosomes. If a transplanted human growth hormone gene were stuck into a mouse chromosome right next to the animal's own growth hormone gene (a feat known as "homologous recombination"), perhaps it would behave like the native gene, sharing the limitations imposed on the mouse gene by neighborhood, structure, and developmental history. In yeast, foreign genes are often inserted into the host chromosomes at the site of the homologous, or similar, natural gene. But mammalian genomes are hundreds of times larger than those of yeast, and transferred genes seldom seem to find their native counterparts. Intruding genes appear to be stuck into mammalian chromosomes at random sites, perhaps by enzymes that normally scurry about repairing breaks and errors in the DNA strands.

The unpredictable and haphazard way in which transplanted genes are usually strung together before being inserted into the chromosomes may also affect their function. Multiple copies of a gene, including broken pieces and bits of DNA attached during cloning, are usually linked head to tail and spliced into the host chromosome as one long chain, perhaps knocking out native genes in the process.

Researchers are already devising new schemes for transferring genes into mammalian cells so that they are spliced in cleanly and perhaps even directed to the site where their natural counterparts sit. Whether either factor or both will allow researchers to mimic completely the natural gene regulation process remains to be seen.

In 1982, the same year one team of scientists engineered runaway growth in mice, another headline-grabbing achievement gave researchers hope that transplanted genes could be made to behave in a normal way. Working in a life form that is a few more evolutionary steps removed from man, a second research team achieved a permanent and equally visible transformation in a creature's physical traits: changing the color of a fruit fly's eyes.

21. *Red-Eyed Flies*

"Where did you come from, baby dear?
Out of the everywhere into the here.
Where did you get those eyes so blue?
Out of the sky as I came through.
　　　　　　　—GEORGE MACDONALD

It doesn't take much to keep a fruit fly's eyes their natural brick red. A single gene called "rosy" working at only 5 percent efficiency can do the trick. And rosy doesn't even have to turn out its enzyme product in the cells of the eye; anywhere in the fly is good enough. However, there are mutant strains of flies without the rosy gene, and they must settle for brown eyes. When a group of researchers from the Carnegie Institution in Baltimore decided to try engineering permanent genetic changes in fruit flies, they chose to transplant a visible trait that would be easy to follow: the rosy gene.

Fruit flies, the workhorses of genetic research for most of the twentieth century, share with bacteria a genetic feature that has not yet been found in mammals—bits of DNA that routinely move about. For a decade scientists have been inserting foreign genes into the movable elements of bacteria, the plasmids, and sending the resulting recombinant DNA molecules back into bacteria for cloning.

Could one of the several mobile genetic elements of the fruit fly—called transposable elements, or transposons—serve a similar function by ferrying foreign genes into fly embryos? That was the question the Carnegie team of Gerald M. Rubin and Allan C. Spradling set out to answer. They chose a transposon called a "P element." This element regularly moves about and inserts itself into different sites on the fly chromosomes (unlike bacterial plasmids, which remain outside the chromosomes as independent loops of DNA).

Rubin and Spradling spliced the rosy gene into a P element and injected the hybrid molecule into embryos of mutant flies that lacked a rosy gene of their own and so were destined to be brown-eyed. For reasons peculiar to flies, the eye color in the first generation of flies was not changed.

Injecting a fly embryo shortly after fertilization is not like injecting a mouse embryo at the single-cell stage and having the foreign gene carried in every cell of the resulting animal. The nucleus of a single-celled fly embryo divides independently several times before the resulting nuclei are compartmentalized into cells—including those that will form the creature's eggs or sperm. Getting the injected DNA into all the nuclei was not practical, so the scientists aimed for the posterior pole of the $\frac{1}{50}$-inch embryo in hopes of getting new genes into the cells that form the germ line. This would assure a complete genetic change in some second-generation flies.

Twenty to 50 percent of the embryos injected that way carried the rosy gene stably integrated in their own germ cells and were able to pass along the trait, producing some red-eyed offspring. The change was permanent. The red-eyed offspring also produced red-eyed descendants.

What makes the rosy gene, without any help from tagged-on viral or animal control switches, work in an apparently normal fashion in a new strain of flies? Researchers believe it has to do with the way the P element settles the gene into its new worksite, splicing it into the chromosome neatly, efficiently, and intact. P elements come with their own enzyme, transposase, for inserting precise segments of DNA into a chromosome. They are not left at the mercy of the cell's own repair enzymes or other mechanisms to become integrated. Genes transferred by other methods are often rearranged and strung together in unpredictable numbers and configurations during the chancy and little understood integration process.

The P element, however, is not targeted to any particular location on the fly chromosomes. It slips in at random sites. It does not seem crucial for the rosy gene to be at its normal chromosomal site, because it can produce its effect successfully from any tissue and without working at top efficiency. Half a dozen other genes transferred by the Carnegie team and by other labs since the rosy success have worked properly, too, including some that normally work only in specific tissues and require precise production levels to accomplish their effect.

Scientists still suspect that some animal genes might need to be at their natural sites to do their jobs properly, however, and if so, they haven't figured out yet how to arrange this with any consistency.

By the time the Carnegie team had announced their results, they had already sent copies of the P element to at least seventy other labs. Researchers were trying to use the new transfer system not only in fruit flies but in creatures ranging from microscopic nematodes to mice. Even if the P element does not work in all these organisms, scientists believe that similar movable genetic elements will eventually be found in most plant and animal species, including man. The speculation is that some of the repetitive lengths of DNA that do not carry instructions for proteins and yet make up the bulk of our genetic endowment are mobile elements, or the remnants of once-mobile elements. With the right enzymes, could some of these be activated again to ferry new genes cleanly into our chromosomes?

With no mammalian P elements in sight yet, some scientists are not waiting to see if the fly element works in higher animals. They have turned instead to a class of viruses—retroviruses—that already travel around within mammalian chromosomes as though they were native transposons. Using all the chopping and splicing tools of modern genetic engineering, researchers are rebuilding these viruses to meet their specifications, turning the tables on them, and domesticating them to serve mankind.

22. *Taming the Retrovirus*

As in all human activities, the rise of techniques can often
act as a lever to implement a change in social practices and
attitudes.
—Hermann J. Muller, 1959

After battling viruses for a century, some scientists are taking a
new tack. Armed with the tools of genetic engineering, they have
set out to tame some and put them to work shuttling genes into
specific target cells. If the modified beasts prove docile and tracta-
ble enough, ironically they may be put to work paying for their
previous mischief by helping to treat human genetic diseases.

The target of the most furious tinkering is the retrovirus, or
RNA tumor virus. It was picked because the whole point of its
natural life cycle is getting into someone else's genes. Other types
of animal viruses, including the monkey virus SV40 which has
been used widely to ferry foreign genes into cells, do not share
this single-minded goal. They are usually content to remain aloof
within the nucleus they invade, using the host cell's machinery to
replicate themselves, but seldom mingling with the local DNA.

Retroviruses work backward. They have no DNA. Their genetic instructions are coded in RNA. Like all viruses, they carry
none of the other requirements for life—such as manufacturing,
processing, or reproductive equipment—just a genetic blueprint
sealed in a protective capsule. For the rest they must depend on
the cell they infect. But a cell is not set up to process genetic
instructions in RNA. The first thing a retrovirus must do when it
gets inside a cell and takes off its protective coat is make a copy of
itself in DNA. To accomplish this, retroviruses carry genetic instructions for a unique enzyme called "reverse transcriptase"
that chemically translates the RNA into DNA. (As we saw in
Chapter 10, this enzyme has proved to be quite useful to genetic
engineers since the early 1970s in making probes for locating and
isolating genes.)

The DNA copy of the virus then slices open a host chromosome, inserts itself, and becomes for all intents and purposes one of the cell's own genes. When the cell begins transcribing this viral DNA sequence into RNA—as it must to get a working copy of any gene for use in protein production—the result is more copies of the RNA virus and the orders for materials needed to make the virus capsules.

"The infected cell doesn't die," says Richard Mulligan of MIT, who is in the forefront of the efforts to tame the retrovirus. "It's very happy. You look at the cell and can't tell it's infected."

Some retroviruses, in fact, have apparently been passed down through generations of host animals for millions of years as part of the genetic legacy of that species. The DNA record also contains evidence of ancient viral promiscuity, a trait that makes the retrovirus look much like the movable genetic elements of fruit flies. The chromosomes of domestic cats carry one viral DNA sequence they must have picked up from ancestral baboons sometime in the past ten million years, and another (feline leukemia virus) that seems to have come from ancient rats. Pigs have also acquired some of their viral heritage from early rodents.

Retroviruses undoubtedly contribute to the genetic plasticity of their host species by occasionally picking up and carrying along animal genes as they move from cell to cell, just as phages pirate bits of bacterial DNA and bestow them on new hosts. It was apparently this kind of pirating that first endowed some retroviruses with tumor-causing abilities. As we saw in Chapter 8, the cancer genes carried by these viruses were captured from normal animal cells. In the hands of a virus, or under certain conditions in their home cells, these genes can trigger the uncontrolled cell growth known as cancer.

To scientists like Mulligan, who had worked with SV40 during the 1970s, the retrovirus began to look like an ideal second-generation vehicle for shuttling genes into living animals as well as into isolated cells. Experiments conducted since the mid-1970s by German scientist Rudolf Jaenisch of the University of Hamburg had shown that when mouse embryos are infected with a retrovirus known as Moloney murine leukemia virus, some of the resulting offspring will incorporate it into all their cells. Like any other foreign gene successfully microinjected into an embryo, the virus passes on to succeeding generations along with the mouse DNA. Presumably, the same fate would await any alien gene researchers could load onto the virus.

"So what we set out to do was go back and use the same sort of strategy that was used with SV40: basically, take out bits of the virus genome, put back in foreign pieces of DNA, and see whether we could manipulate the virus to do what we wanted. That is, get genes into chromosomes very quickly," Mulligan says.

But there was a catch: You cannot use recombinant DNA techniques to tag bits of DNA onto a string of RNA.

"The basic conceptual problem is the fact that the genome is in RNA. The chemistry isn't available to manipulate RNA the way we can with DNA," Mulligan notes. But he and other researchers came up with a solution, and their elaborate tinkering reflects the enormous powers genetic engineering has given us to modify microbial life forms at will.

First, researchers took advantage of the retrovirus's own strategy of copying its RNA into DNA inside an infected cell. They isolated this DNA and cloned it. Would these man-made DNA copies of the virus work in a cell? The answer was yes, but getting them into cells was not easy.

The DNA copies were not encapsulated in a protein coat and so were noninfectious. To get them into cells, researchers had to rely on inefficient DNA transformation techniques: bathing the cells in the viral DNA, then searching for the one in ten thousand or one hundred thousand that took in the foreign genes. In the cells that did incorporate the viral copies, however, the DNA successfully stuck itself into a chromosome and began ordering up new RNA virus particles and coat proteins. These second-generation viruses were efficiently infectious, identical to the ones that would have been made had an RNA virus infected the cell.

If the man-made viral DNA could get itself into a chromosome and generate its own products, Mulligan figured it could probably do the same for any foreign gene that was hooked to it. To make room to attach the foreign gene, however, his team had to gut most of the midsection of the virus. The resulting hybrid DNA molecule included a foreign gene and an incomplete stretch of viral DNA. When this hybrid DNA was transferred into mouse cells, it began ordering products just as the full viral DNA copy had done. This time, however, the products included those coded for by the foreign gene. For the virus this scheme was a dead end. To make room for the passenger, Mulligan's team had had to strip away the viral genes that carried the instructions for encapsulating its offspring and thus making a second generation of infectious viruses.

At that stage the system wasn't much of an improvement over what had been achieved using SV40 and other viruses: a foreign gene attached to a strong viral control switch, all packaged in a hybrid molecule that had no infectious powers. The researchers still had to rely on inefficient DNA transformation techniques to get the hybrid into cells, just as they had with other viral hybrids. The retrovirus system had only one advantage at that point—the DNA package tended to nestle into the host chromosomes more efficiently than other viral hybrids.

But Mulligan wanted to add another stage to the system and get the second-generation copies of the defective virus and its passenger gene packaged into infectious particles. Making use of a trick from past viral research, he teamed up another retrovirus with the hybrid as a helper. Once they were both inside a cell, the helper ordered all the proteins needed to package both its own and the hybrid molecule's RNA offspring. The helper thus turned the cell into a factory for infectious hybrid molecules. The second-generation hybrids were still defective—chunks of the viral genes were missing—but this did not suppress their infectious powers once they were sealed in a viral coat.

"One extraordinary thing about retroviruses is that once a defective RNA is packaged into a particle, it can not only infect a cell, but it also contains all the genetic machinery necessary to get it into the chromosome efficiently. The enzymes and various other molecules necessary for the integration process are actually packaged into the coat as opposed to being synthesized from the instructions in the viral genome," Mulligan notes.

The result was a benign Trojan Horse, a recombinant molecule disguised in an infectious viral coat that allowed it to slip efficiently into a cell and insert a foreign gene into the chromosome. At this stage the packaged hybrid had all the invasive powers of a natural virus. No helper virus was needed to get it into the cell. Once inside, however, it was stuck again. Without a helper the viral genes could not package a third generation of infectious particles. For the virus, and its genetic passenger, it was a dead end.

"But, in fact, that's exactly what you want in a gene transfer system," Mulligan affirms. "That is, to be able to put the foreign gene into this cell very efficiently, but then to see no evidence of viral gene expression or infection." The viral genes, and more important, the foreign gene the team wanted to insert, are now stuck in place. If a live mouse, (or a person) were to receive new genes this way, it would not be harboring an active virus infection that could be spread (foreign genes included) to others.

Another clever round of tinkering eliminated the need to separate all the helper viruses's progeny from the recombinant particles to ensure there would be no infectious third generation. "We went back and engineered the helper virus to be defective in a very specific way—in the ability to package itself into a particle. This virus is still able to make all the viral proteins necessary for packaging the recombinant molecules, but it can't package its own genome."

A second bit of engineering on the helper virus took care of another drawback in the original system. The team started with retroviruses that only infect mice, not monkey, human, or other cells. But another mouse retrovirus—a leukemia virus isolated from a wild mouse species in California—can carry on its business in all kinds of cells. The scientists didn't need to start over again and tame this new virus as they had tamed the first one. The viral coat is the password that determines which cells the parasite can enter. Mulligan's team simply stole the coat for their own defective helper virus by taking the coat gene from the new retrovirus and splicing it into the helper to replace its own coat gene.

"And what we end up with is now what we would call the ultimate gene transfer system," he says. "These viruses [the recombinants packaged by the defective helper in a coat from a third virus] will infect almost anything under the sun and then be a dead end, which is exactly what you'd want."

The complete system for getting any alien gene into an animal cell is as follows: First, the gene is spliced into the tamed and defective retrovirus and inserted into a few cells using DNA transformation techniques. These transformed cells are also infected with the engineered helper virus. The infected and transformed cells obligingly churn out copies of both the helper virus and the recombinant molecule, but the helper-virus proteins package only the recombinants into infectious particles.

"The [recombinant] virus actually buds into the culture fluid of the medium that's covering the cells," Mulligan explains. "So you take the media and now you have an infectious virus stock. And you put that on other cells, and lo and behold . . . it's remarkable just to see what happens, especially if you've been using the DNA transformation method. If you take a milliliter of the culture fluid and put it on cells, in contrast to getting something like fifty colonies with DNA transformation, the whole plate is transformed. When you put the selective agent on [to kill off the cells that have not taken in foreign genes and put them to work], it doesn't do anything. All the cells live because they all get transformed.

"So it's really very powerful. And then when you look at these cells to see whether they produce any retrovirus—which they would if they had a helper—they don't. So it's a perfect way to introduce genes into cells."

This tamed and neutered retrovirus offers the most efficient method yet devised for getting new genes into virtually every cell in a target population—without eventually killing the cells or causing disease as viruses like SV40 do. As a mobile recombinant particle the retrovirus also performs a function similar to that done by the P elements in a fruit fly: It integrates foreign genes cleanly and efficiently into the chromosomes of animal cells.

But would the virus-delivered genes, like those in the fly, be regulated in a normal way? Mulligan teamed up with Jaenisch in Germany to test the system in embryos, to see if infectious hybrid genes could go to work and make "new mice." Jaenisch had been infecting mouse embryos with whole retroviruses for a decade. He had found that retroviruses injected in the very early embryo were yoked with molecular baggage called methyl groups, and for that or other unknown reasons, they were infrequently activated in the cells of the adult mice.

However, Jaenisch had developed dozens of strains of mice that passed on these freeloading genetic sequences from generation to generation in the chromosomes of the sperm and egg cells. Each strain carried the virus on a different chromosome, confirming that the virus, like the P element, does not seem to home in on a specific site. Despite initial methylation the viral genes were later demethylated and turned on in some lines. Virus particles were actively produced in the adult animals at various stages of development and in various tissues, depending on where the virus had settled in the chromosomes of each line of mice.

This varied and unpredictable expression was not seen in fruit flies when the mobile P element was used to ferry in new genes. Those fly genes seemed to produce their products at the right time and in the right amount no matter what part of the chromosome they landed in.

The tests with the whole retrovirus, however, may not provide an accurate picture of how genes ferried in by a defective virus will be expressed. After all, there is no way for a mouse cell to express viral genes "normally." What tissue should a mouse make viruses in, and in what amount and when? Perhaps the early embryo slaps methyl groups on the viral genes and turns them off for the same reason. "The embryo doesn't know what to do with a retrovirus," Mulligan speculates. "It's not a normal gene. The embryo doesn't know whether it should be on or off."

In 1983 Jaenisch and Mulligan used the retrovirus vehicle for the first time to ferry a new gene into mouse embryos. (The virus does not have to be microinjected into the nucleus the way naked DNA does. The two-to-four-cell embryos can just be dropped onto a plate of virus-producing cells and about half will become infected.) The first test was designed to confirm that the system worked, that the foreign gene would be integrated into all the cells of the mouse and passed on to the next generation. It was. But the foreign gene the retrovirus carried was a bacterial enzyme gene that the mouse would have no reason to know what to do with.

The hope for future experiments is that when one of the team's hobbled and domesticated retroviruses arrives carrying a gene the mouse embryo recognizes, the mouse will know how to turn it on and off appropriately no matter where it lands, just as the fly apparently does with its P-element-delivered genes.

The same hope holds for transferring new genes into isolated cells such as bone marrow, then implanting these cells into a living animal—the strategy Martin Cline used in his 1980 attempt at gene therapy in humans. But Cline tried to transfer hemoglobin genes using DNA transformation techniques that at best get new genes into one of every ten thousand cells. His targets, the stem cells that create new generations of blood cells, are present in the bone marrow at about the same ratio—one in ten thousand. Therefore, the chances were slim that enough new genes could be transferred into a significant number of stem cells using that method to make a change in the patient's condition.

"In theory we don't have that problem, because if we just dump on enough virus, every cell in the marrow population will get the genes stably incorporated," Mulligan says. White blood cells and other marrow components, however, are not supposed to make hemoglobin. It is hoped that not every cell that receives a retrovirus-carried gene will activate that gene, and that no matter where the gene lands in the chromosomes of any particular cell, the cell will know exactly what to do with it.

The first bone marrow transplants that are being done using the new system, as in the first embryo infections, are designed to prove only that the retrovirus can get new genes into a living animal efficiently. A second round of tests with genes like hemoglobin will eventually provide more interesting answers.

A lot of work remains before all the questions about gene control raised by this second generation of transfer techniques are answered.

"Now we can ask for the first time, if we put the gene into all the blood cells, will they be expressed in every cell or just in the appropriate blood cells?" Mulligan says. "No one has been able to ask these questions before.

"We're now going back to what I would call at least superficially the boring experiments: just putting globin genes back into fibroblast [skin] cells and other genes into environments where they shouldn't work. This is what everyone did initially because everyone was excited: 'Look, we can get genes into cells. Let's put a globin gene into a fibroblast cell and hope it works.' And everyone hoped it would work for a while, and then it did. Then everyone realized they really should hope it doesn't work in those cells. So now what we want to do is to go back and repeat some of those kinds of experiments and maybe show with the retrovirus system that the genes don't work in the wrong kind of cells."

The tinkering with the retrovirus continues along with the gene transfer experiments. For the future Mulligan envisions using "envelope" genes pirated from many different kinds of viruses, including influenza viruses, to coat and aim the tamed viral carriers to specific target cells from various species. One gene being tested because of its wide host range—"it infects all kinds of cells from man to mosquito"—is the coat gene from a vesicular stomatitis virus.

Even more ambitious is the possibility of sending the tamed retroviruses into living creatures and programming them to home in only on specific tissues where new genes are needed— bone marrow, nerves, muscles, kidneys. One possible targeting scheme is to attach highly specific monoclonal antibodies with an affinity for a particular tissue type to the virus particles to see if that will get the viruses into the desired cells.

"The viruses have a lot to offer for gene transfer just because they've evolved to do very specific things and do them well. That's why they are where they are. That's why they still exist at this time in evolution. And it's up to us to be able to exploit those particular properties—take a property from this virus and a property from that virus and use them to our advantage," concludes Mulligan with a confidence undreamed of a decade ago.

Whether tamed retroviruses will be the gene transfer system that brings the first successes in human therapy is still unknown. Scientists and physicians in the 1980s envision a broad range of possible uses for this and other transfer systems to put "good" genes into patients who need them. Other researchers are looking at ways to send signals to correct or bypass the "bad" genes the cells already carry.

V
GENETIC FUTURES

23. *Healing with Genes*

Not only the utopians, but tacitly all men recognize a
plasticity or multipotentiality in man—that individual life is
not fully determined. Yet the biologist knows that genetic
constitution provides the constraints within which man's
whole capacity must develop and which limit the scope and
possibility of training, hygiene and medicine. Small wonder
that among the utopians those with biological inclinations
have outlined means to escape the lottery of genetics . . .
—ROLLIN D. HOTCHKISS, 1965

From the standpoints of futuristic gadgetry and high drama, gene
transplants in humans will probably be something of a bust. Patients won't be facing anything like the fearsome technology and
trauma of an artificial heart, the devastating toxicity of a powerful
new cancer drug, or the painful struggle to prevent rejection of a
transplanted kidney. Gene therapy, for all its profound implications, will look rather prosaic in practice.

The first wave of patients, like the thalassemia victims Martin
Cline attempted to treat with new genes in 1980, will probably be
very familiar with hospitals and painful medical procedures. Their
doctors will have tried all the conventional treatments, to no
avail. Their futures will be quite limited. New genes, for them,
will be a last hope.

Using only a local anaesthetic, physicians will insert a biopsy
needle into a patient's hip to remove a sample of bone marrow, or
perhaps into the liver to retrieve a small section of tissue. Back in
the lab the marrow or the liver cells will be infected with hybrid
virus particles or some other kind of gene carrier, each one loaded
with a copy of the normal human gene the patient lacks. Later in
the day the marrow cells will be reinfused into the patient's bloodstream to find their way to a niche in the bone marrow; if liver
cells are used, they may be surgically reimplanted.

As procedures improve and methods are developed for targeting the viruses or other gene carriers to specific cells, the pro-

cedural steps of tissue removal and return will be dropped. New genes will be injected directly into the patient just like vaccines.

After the implant will come the wait. The invisible invasion, silent and unfelt, may take weeks or months to make an impact. If the new genes work, however, and the cells that carry them proliferate, they will begin churning out supplies of an enzyme or blood protein the patient is missing. Tests will show the red blood cell count rising, stored wastes being flushed from tissues, body defenses coming to life, or nutrient processing returning to normal. The result should be a complete cure of the patient's affliction.

(As in all medical experimentation, some of the first patients to try a new therapy may already be irreversibly harmed by their diseases—thalassemia victims whose hearts are crippled by iron deposits, or victims of storage diseases whose kidneys or nervous systems are damaged. If the first trials show the therapy is effective and not unduly risky, it can then be offered to others suffering from genetic diseases before any permanent damage occurs.)

The decade of the 1980s will see a substantial rise in gene transplant experiments in live animals and also the first careful efforts to use gene therapy to treat human disease. By the 1990s gene therapy may have achieved the status that test-tube baby technology now enjoys—not yet an everyday occurrence but already too common to command headlines or a slot on the network news.

Gene therapy will not be a panacea, of course. No medical therapy is. In an increasing number of diseases, patients will be able to lead normal lives with regular infusions of gene products churned out by engineered bacteria. Children with inherited deficiencies in human growth hormone, for example, will not need a permanent supply of new genes in their cells if a limited round of hormone injections can allow them to grow to a normal height. As scientists get to the root of more disorders, they will find numerous instances in which infusions of a drug or biochemical can block or reverse the harmful actions of a "bad" gene product—perhaps, for instance, the unknown product that triggers the destruction of brain cells in Huntington's chorea victims. Diseases that involve both defective genes and environmental triggering agents may be prevented in many cases by blocking the action of the external agents. For individuals with many other genetic diseases, however, an infusion of new genes will provide the best or perhaps only hope for a normal life.

In 1980 physician and molecular biologist W. French Anderson and bioethicist John C. Fletcher of the National Institutes of Health recommended three criteria that scientists working in animals should satisfy before attempting further gene therapy in humans. First, they must be able to deliver genes to the right target tissues and ensure that the genes stay and work only in those cells (or that they do not hurt anything if they wander or express themselves in some inappropriate places). Second, the genes should be appropriately regulated so that they make enough of their product to correct the disease but not so much that they cause harm. And third, scientists must show that the intruding genes do not foul up cell processes in a way that harms the animal. Other physicians believe the criteria should be more flexible for the first trials in patients with no other hope.

Decisions about when to attempt gene therapy will probably be made at the level of medical center and university review boards as individual research teams seek permission to try their procedures on patients with specific types of disorders.

The first candidates for gene therapy trials will most likely be adults or children with well-understood diseases caused by defects in single genes—defects that might be cured by a transfer of new genes into an accessible and self-renewing tissue like bone marrow.

Hemoglobin disorders such as sickle cell anemia and the thalassemias usually lead the list of first targets for human gene transplants because the defects that cause them are understood at the molecular and in some cases even at the atomic level. Also, the blood-forming substance, the bone marrow, is easily accessible. But as we know, hemoglobin production is a tricky family affair. Getting all the globin genes working together in the right combinations at the proper stage of life and producing massive but coordinated amounts of proteins for use in hemoglobin assembly requires some touchy regulation. It may be easier to work out the bugs first with a gene for a single enzyme that is produced continuously throughout life and requires no assembly.

One possible candidate for early gene therapy trials is severe combined immunodeficiency disease, which is caused by the lack of either of two enzymes—adenosine deaminase or purine nucleoside phosphorylase. Unlike hemoglobin, these enzymes must be produced at low, constant levels. Infants born with this disorder, as we saw in Chapter 9, lack both the components required for normal body defenses, cellular and humoral immunity, and often die from common infections in early childhood. The cells that give rise to both types of immune components reside in the marrow. Since 1968 a number of patients with combined immunodeficiency diseases have been cured by bone marrow transplants, but only a small percentage of victims of this disorder have suitable marrow donors. Gene transplants would overcome this problem.

Any genetic disease where enzyme replacement therapy has been considered would also be a candidate for gene transplant. Some of the approximately thirty storage disorders, such as Fabry's disease or the milder form of Gaucher's disease in which the nervous system is not affected, have been mentioned as possible early targets. Each of these disorders is caused by the lack of a particular enzyme that cells need to digest specific lipids or fatty substances. Without a continuous supply of the enzyme, the fats accumulate and damage cells.

Researchers have claimed a few modest successes in temporarily lowering tissue lipid levels with infusions of the missing enzyme, but to help the patient, the enzyme must be supplied to the right tissues throughout life. Gene therapy could make this possible for the first time. (Infants with the most widely known of these disorders, Tay-Sachs disease, probably would need to be treated in the womb, since fatty substances begin building up and damaging their brain cells before birth.)

Another group of likely targets is the inherited enzyme deficiencies that prevent the proper handling of amino acids in the proteins we eat—PKU, homocystinuria, maple syrup urine disease, and histidinemia. The retardation associated with PKU can be prevented if the disease is diagnosed at birth and the child is kept on a strict diet until his brain development is complete. But diet is no cure. If a woman with PKU goes back to normal protein intake after her own brain development is assured, high levels of the amino acid phenylalanine in her blood during pregnancy can irreversibly damage the brain of any baby she bears.

People who inherit one gene for the dominant disorder familial hypercholesterolemia have only one half the normal number of receptors for low density lipoprotein (LDL), the cholesterol-carrying molecules in the blood. Receptors are needed to grab the LDL and cholesterol, remove them from the blood, and take them into cells for processing. Having too few receptors results in high blood-cholesterol levels and the risk of premature heart attack. This risk can usually be controlled by a low cholesterol diet, a prudent life-style, and drugs.

The few individuals who inherit two genes for the disease, however, usually produce no LDL receptors at all. Their blood cannot get rid of its excess cholesterol, and, what is worse, their liver cells do not get the signal that the blood is overloaded. The liver keeps making more cholesterol and pumping it into the blood. Even a cholesterol-free diet is of no use, and these hypercholesterolemia victims may face a heart attack before they enter kindergarten. What would help is a dose of genes to produce LDL receptors.

A number of investigators would like to try putting new genes for the enzyme HGPRT into the marrow of infants with Lesch-Nyhan syndrome. Researchers at the University of California at San Diego and Baylor College of Medicine in Houston have both isolated the human HGPRT gene, teamed it up with retrovirus controls, inserted it into defective cells in a lab dish, and shown that the gene functions there. But the most devastating impacts of this disease are the brain and behavioral effects—retardation and compulsive self-mutilation—and no one has learned yet how to direct new genes to the brain or central nervous system. Some researchers are hoping, however, that a supply of the enzyme produced in the marrow and released into the blood might protect the brain.

Transplanted genes will not necessarily have to work at full efficiency to cure many of these disorders. Carriers who inherit only one gene for a recessive disease usually get along fine with only one "good" gene and 50 percent of the normal supply of that gene's product.

Researchers have not even begun to consider how to alter traits or diseases that involve the interaction of multiple genes. But one hope is that in some of these conditions, the identification and infusion of a single key gene may tip the balance and ease or cure the patient's symptoms.

For finicky genes such as the globins, some researchers think it may turn out to be easier to "correct" the patient's own defective genes than to insert new ones and get them integrated and working properly. This has been called "genetic surgery," and no one has accomplished it on a gene that is still inside a cell. But researchers do know how to make specific changes in DNA sequences—deleting whole regions or changing single molecules—in a test tube.

The process is called "oligonucleotide directed mutagenesis." It makes use of synthetic strands of DNA, man-made gene fragments that are only fifteen or twenty chemical units long and which have been assembled by hand or by gene machine. Each fragment matches the natural sequence except at the site of the desired mutation, the reordered stretch of genetic instructions that the researcher wants to build into a gene. When the short synthetic sequence is hybridized or paired with the much longer natural gene, the two form a section of double-stranded DNA matched up like a zipper with a few teeth missing. The researcher then adds enzymes that use the natural gene as a template to complete the synthetic fragment, assembling it into a full-length strand complementary to the natural one. (Of course, a researcher could synthesize the full-length mutant gene himself, but this is time-consuming. Just assembling fifteen to twenty units of DNA takes all day.)

When this mismatched double strand of DNA is inserted into bacteria for cloning, the bacteria treat it like a normal gene, unzipping the two strands and using each as a template for producing a complementary DNA sequence. The new DNA sequences formed from the strand with the mutation will all carry the mutation. The result is a supply of mutant genes.

Some scientists envision a day when we may be able to correct a defective natural gene in a cell by such a method, by sending in a repair sequence that would home in on the faulty gene and impose new instructions on future generations of DNA passed on to daughter cells.

In April 1982 a group of researchers reported they had taken a different tack in trying to "correct" a defective globin gene. Their method actually requires no changes in the globin gene itself. The idea might be called "two wrongs make a right." It involves sending another defective gene into the cell to misread the instructions of the faulty globin gene and to produce the right protein in spite of bad information.

The experiment was performed by a team led by Yuet Wai Kan of the University of California at San Francisco. Kan had taken defective human globin genes (actually mRNA or working copies of the gene) found in some severe forms of beta-thalassemia and injected them into frog eggs. This defective mRNA carries in its alphabetic code a premature stop signal. Where the code should read AAG for the amino acid lysine, a mutation has converted it to UAG, a signal to the protein-manufacturing machinery to stop. With this stop signal in the wrong place, the beta-globin proteins made from these instructions are abnormally short and cannot be used to assemble effective hemoglobin.

Instead of trying to put a new gene with the right beta-globin instructions into the cell, Kan visualized changing the way the defective instructions get read. As we saw in Chapter 20, a molecule called transfer RNA or tRNA works the protein assembly line in the cell, reading the genetic instructions in mRNA and figuring out where to start, what amino acid to add next, and when to stop. What Kan envisioned was a mutant tRNA gene whose product—a suppressor tRNA—would overrule and suppress the improper stop signal in the globin instructions and turn out complete protein chains.

The team used recombinant DNA techniques to produce such a mutant gene. Its tRNA was designed to misread UAG as AAG. When the suppressor tRNA was injected into the frog eggs containing the defective beta-globin instructions, the scheme worked. The tRNA came to the UAG part of the message and put a lysine on the chain instead of stopping. The result was the production of normal beta globin.

The suppressor tRNA was not totally effective. At best, it properly misread the stop signal 50 percent of the time, sometimes only 20 percent. But 20 percent of normal beta-globin production might be enough to reduce or eliminate a patient's dependence on blood transfusions.

The next step toward the possible treatment of patients with such a system is to determine whether mammalian blood-forming cells can be altered this way without side effects, without suppressing some stop signal that shouldn't be unstopped, at least at significant levels. Every gene in the body, in fact in all biological systems from viruses to whales, makes use of one or another of three chemical stop signals that bring a halt to protein production. These signals are called amber (UAG), ocher (UAA), and UGA.

Kan used a suppressor that overruled an amber stop signal because this is the premature stop that is involved in some forms of beta-thalassemia. Another research team led by Phillip A. Sharp of MIT began working with these suppressors because amber stop signals seem to be relatively rare. Sharp examined the sequences of known viral and animal genes and concluded that not many proteins are "amber-terminated"—that is, not many use the code UAG as a stop signal. Sharp's work is aimed at probing the genes of viruses, not at developing gene therapy for humans, but the team's experiments lend support to the notion that an amber suppressor could be used in humans without causing harm to normal genes.

Sharp's team created an amber suppressor that "misreads" UAG and adds tyrosine to the protein chain instead of lysine as does Kan's suppressor tRNA. Mario R. Capecchi at the University of Utah in Salt Lake City used microinjection to create a number of animal cell lines carrying active copies of the suppressor gene. The cells grow normally, apparently unharmed by the activities of the new tRNA.

"This demonstrates that amber suppressors are not highly deleterious and that Kan's assumption might have been right," Sharp says. "That you might get enough suppression activity to help the patient without killing all the cells in the lineage. I was shocked to find that those cell lines grew normally, with no significant slowing of rates of growth, despite having a functional suppressor." Sharp's team is continuing to study the cells that carry the suppressor tRNA to look for effects on specific genes that have amber stops and to quantify how efficient this suppressor is. Other groups are putting the suppressor into mouse embryos, fruit flies, nematode cells, and yeast to see if it causes any malfunctions.

Sharp's team is also making an ocher suppressor to determine just how deleterious it will be when it is inserted into cells. More than one half the genes in a cell end with an ocher stop, so the ocher suppressor is not likely to have any clinical use. "But it is possible that the cell has enough slop and degradation systems that it could tolerate ten or twenty percent of these proteins being made aberrantly and being destroyed," he says.

All the schemes for putting new genes into humans must depend on findings from animal studies that harmful effects are rare, or at least are much less of a hazard to the patient than the disease he or she inherited. What is the likelihood that an intruding gene will bump into something important and break a chromosome, cause a mutation in a cancer gene that will lead to malignancy years later, or foul up some critical native gene and give the patient a new genetic disese? These are possibilities that must be considered.

Until researchers learn how to direct transplanted human genes to specific sites on the chromosomes, foreign genes will settle into their new homes at random locations. When a quantity of bone marrow or other tissue is exposed to new genes, the genes will probably take up residence at a different site in every cell. If a new gene is too disruptive in any cell, the cell will die. A few dead cells in a large population, or a cell or two with malfunctioning genes or broken chromosomes might have no impact on the patient. One cell that turns malignant, however, might lead to cancer.

Since 99 percent of the DNA in every cell is garble and filler of no known function, some scientists believe the possibility of a transplanted gene bumping into and damaging something important is less likely than it might sound.

"There are fifty thousand genes somewhere in there, but there's enough DNA for a million genes," says Jon Gordon of the Mount Sinai School of Medicine. "So there's a lot of DNA in there that's not functioning in any obvious way. Some people would say statistically your chances are twenty to one of getting it [a transplanted gene] into a nongene region . . . And then there's the fact that in any differentiated cell type like the bone marrow, of fifty thousand genes which are capable of functioning, only some are. The albumin gene that would be functioning in the liver is not active in bone marrow. So let's cut that fifty thousand down to ten thousand functioning genes. Now out of a million genes worth of DNA, you only have ten thousand gene sequences that are in danger when you put a gene in. So some would say statistically it's worth it. I have no answer. I think we need a lot more experiments to get the answer."

Other scientists suggest that as better cell culture techniques are developed, cells can be infected with new genes in a lab dish and grown there for days or weeks before being returned to the patient. Technicians will be able to perform biochemical tests to make sure the genes are working and watch for malfunctions or malignant changes before reimplanting the cells.

We have been talking so far about making genetic changes in the body cells of individuals, changes that will die with the patient and not affect his or her offspring. If transplanting new genes into human embryos is ever attempted—and most people in the field do not feel this will be an option in the foreseeable future—the hazards involved in the procedure will be a bigger problem. A few cells with mutant genes or broken chromosomes among trillions will usually have no impact on the creature that carries them. But a fertilized egg is one cell, and if it is damaged by the entry of a foreign gene, that damage will be carried in every cell of the developing individual.

In one of Gordon's mouse lines created by embryo injection, the implanted interferon gene has never functioned; but during its entry it managed to knock off a piece of chromosome 2 which later attached itself to the end of chromosome 12. Males in this lineage are sterile. In one of Rudolf Jaenisch's mouse lines at the University of Hamburg, the mice carry a retrovirus inserted into a particular site on their chromosomes. The mice in this lineage produce small litters because the fetuses that inherit the virus from both parents die on the fifteenth day of gestation, which indicates the foreign DNA has interrupted some crucial native gene sequence involved in that stage of normal development.

As we saw in Chapter 19, there is no way presently to detect genetic defects in a fertilized egg or an early embryo. Even in diseases with the worst odds—dominant disorders in which the offspring have a fifty-fifty chance of inheriting the defective gene—it would be hard to justify injecting corrective genes into a possibly healthy embryo until the survival rate of injected embryos improves and the sorts of genetic damage found by Gordon, Jaenisch, and other researchers can be better controlled. (This also holds true for any "improvements" people would like to squirt into otherwise healthy embryos.)

Bernard D. Davis of Harvard has noted that "there is a practical consideration that will deter responsible investigators from altering human embryos for a long time to come: the need for virtually perfect reliability. In somatic [body] cell therapy a 50 percent cure rate would be a triumph, but manipulations of embryo cells that damaged even one child in a thousand would be intolerable."

It may be possible in some cases to manipulate human genes without actually putting new genes into people. Present drug therapies are aimed at correcting or compensating for genetic effects, not at altering the genes themselves. Scientists would like to be able to give a patient a drug or chemical that would actually get into the proper cells and alter the functioning of selected genes. The first attempt to do this came in 1982, and the results were promising.

During that year two research groups treated a half dozen beta-thalassemia and sickle cell anemia patients by using a drug called 5-azacytidine to turn on genes that had been shut down since birth. The treatments did help to correct the patients' severe anemia temporarily, and they marked the first time scientists had been able to alter deliberately the activity of genes in human beings. But 5-azacytidine is not necessarily selective about which inactive genes it turns on, and the treatment is not yet recom-mended for general use.

Still, in an editorial in the *New England Journal of Medicine* Edward Benz, Jr., of Yale noted that the work demonstrates molecular biology "has come to the bedside."

Adult hemoglobin is made up of two types of protein chains, alpha globin and beta globin. During fetal life, however, the beta-globin genes are normally switched off and the fetus produces another type of globin chains—gamma—to match up with the alpha chains and form fetal hemoglobin. Just before birth the genes swap status: Beta is switched on and gamma shuts down. It is the beta gene that is defective in both beta-thalassemia and sickle cell anemia. Researchers had proposed years ago that turning the gamma-globin genes back on might cure these disorders.

"The ability to turn off the synthesis of beta-globin chains and maintain the 'switch' for gamma chains in the 'on' position would make sickle cell anemia and beta-thalassemia harmless experiments of nature rather than progressive and usually lethal disorders," geneticist Leon E. Rosenberg of Yale noted in 1979. There was no question the gamma chains could do the job if they were available. Patients with one rare inherited disorder (hereditary persistence of fetal hemoglobin, or HPFH) produce only fetal hemoglobin all their lives and apparently are not harmed in any way by it. The problem was how to turn on the dormant gene.

The drug 5-azacytidine was already being used in the treatment of leukemia and seemed to offer hope. When cells in a lab dish are dosed with the drug, a wide range of dormant genes are sometimes activated at low levels. In 1981 researchers, led by Joseph DeSimone and Paul Heller at the University of Illinois, tried 5-azacytidine on baboons and found it increased their production of fetal hemoglobin. (When baboons are under stress, they normally make a certain level of fetal hemoglobin along with the adult variety.)

These results encouraged a group at the National Heart, Lung and Blood Institute led by Timothy Ley and Arthur Nienhuis to team up with DeSimone and Heller to try the drug in humans. The team treated three beta-thalassemia and two sickle cell anemia patients, all of them severely debilitated by their diseases. Another group at Johns Hopkins in Baltimore, working independently, treated one sickle cell anemia patient.

Both groups reported that infusion of the drug over three days to a week dramatically but temporarily increased fetal hemoglobin production. Red blood cell counts increased significantly. In sickle cell patients, as fetal hemoglobin increased, the production of sickle hemoglobin decreased. These impacts on the blood were promising, but the tests were too brief to demonstrate any therapeutic benefits to the patients. Future studies will have to determine whether this drug, or others, can end the need for transfusions in thalassemia victims or prevent sickle cell crises.

Besides the questions about therapeutic benefits, several others remain. One concerns the toxicity of the drug over time. Both research groups administered lower doses than are usually given to cancer patients and saw no side effects. But in leukemia sufferers, 5-azacytidine damps the production of white blood cells and platelets, leaving the patients vulnerable to infections and bleeding disorders. Like many other chemotherapy drugs, it also causes nausea and vomiting. And in mice at least, the drug itself can cause cancer.

Another question is whether the drug switches on other genes that it might not be desirable to have activated in the wrong tissues at the wrong times. Many lab scientists were taken aback by the clinical trials, just as they had been by earlier gene transplant attempts, because of the rather profound and unspecific gene-activating effects the drug has on cells in a lab dish.

(In 1965 Rollin D. Hotchkiss of Rockefeller University, speaking of future genetic manipulations in man, cautioned: "In attempting to have the elements so mixed as to produce the noblest Roman of all, one might unwittingly produce a Dogberry, a Caliban, or quite conceivably a mosaic monster. The presumably selective mutagens and the agents aimed at modification of controllers and repressors [of gene activity] are among the measures most likely to give unexpected results here.")

Experiments with 5-azacytidine should provide researchers with new information about how genes are normally switched on or off, which may help in designing new gene-activating or gene-modifying drugs that are less toxic and more specific in their targets. Earlier work had shown that when the fetal gamma-globin genes are suppressed, they are covered with the chemical units known as methyl groups. When these genes are active during fetal life, they are not covered by the methyl groups. In cell culture studies and in tests of patients' DNA, gamma-globin genes were demethylated after treatment with 5-azacytidine. But there is still no proof to confirm whether the removal of methyl groups is a cause or a consequence of gene activation. Demethylation may be only one of several steps necessary for the normal expression of most genes.

Other therapeutic visions prompted by advances in genetics range from modifying the self barriers that limit organ transplants to altering the specialization of cells that prevents the regeneration of most injured human tissues.

To keep patients' bodies from rejecting transplanted organs, physicians often must blast their entire immune systems with powerful drugs that leave them virtually defenseless against infection. Eventually physicians would like to be able to alter the immune system selectively, depleting certain cell populations or adding components to the system that would modulate specific immune responses.

Other researchers speculate about a day when they may be able to change the molecular monogram of a tissue or organ before transplant, marking its cells with the same combination of self tags or HLA antigens borne by the cells of the patient receiving the transplant and thus preventing rejection. Transfers of HLA genes into people might also be used someday to improve resistance to viral infections or to lessen vulnerability to autoimmune diseases such as multiple sclerosis.

Already genes for some HLA markers have been introduced into cultured cells, and the antigens produced from them are displayed on the cell surfaces. This not only alters the tissue type, but it also changes the interactions between those cells and immune system components such as the killer T cells that protect the body against viral infections.

Other visions go beyond alteration of the immune system to extending the human life-span. By altering or adding to the genes of the HLA region, some researchers believe it might be possible to endow the cells of an individual with higher levels of DNA repair and internal protective mechanisms. This might increase our ability to withstand exposure to toxic chemicals, radiation, and other environmental hazards, reducing the risk of cancer and other ailments and perhaps even staving off the effects of aging.

Future gene therapy will not have to be limited to the transplant of natural human genes. Synthetic genes for products nature has not supplied us with are already being constructed and inserted into bacteria to create novel drugs. For instance, both hybrid natural interferon genes and synthetic interferon genes are being assembled in an effort to optimize the antiviral or anticancer properties found in various types of interferon.

A more distant and complex vision involves human control of the processes of cell differentiation that guide the growth of an entire individual from a single fertilized egg, shaping limbs and internal structures and laying the foundations for biochemical cycles and behavioral possibilities. The vision is not recent:

"With a more complete understanding of the functioning and regulation of gene activity in development and differentiation, these processes may be more efficiently controlled and regulated, not only to exclude structural or metabolic errors in the developing organism but also to produce better organisms," geneticist Edward L. Tatum predicted in his 1958 Nobel Prize lecture.

It was changes in the timing and course of development that allowed man to diverge from his ape cousins. Some visionaries foresee the possibility of another round of major changes in human shape, structure, brain size, and temperament, this time self-directed. If the remaking of man is ever possible, there is no lack of suggestions for, or controversy over, how we might want to redraw the design.

24. Genetic Twilight?

We cannot expect nature to start improving our innate abilities once more. The usual fate of a species in the past has not been progress, but extermination, very often after deteriorating slowly through long periods. The animals and plants alive to-day are the descendants of the few species which have escaped this fate. There is no reason to suppose that man will escape it unless he makes an effort to do so. And we do not at present know how to make that effort. Doubtless complete idiots should be prevented from breeding, but the effort to eliminate all sorts of "unfit" human types is a very much more dubious proposition . . . Many of the "unfit" are unfit for society as it is to-day, but that is often society's fault. The attempt to prevent them from breeding really involves the appalling assumption that society as at present constituted is perfect, and that our only task is to fit man to it.

—J.B.S. HALDANE, 1933

Isolated in a remote river valley in the Arizona desert for more than two thousand years, the Pima Indians learned to survive harsh cycles of feast and famine. The River People, as they still call themselves, depended on the Gila River to fill the irrigation canals they dug and supply precious water to their corn. Every five years or so, drought struck and the crops failed.

That life is gone now. Dams built upriver stopped the flow of water a century ago and turned the wooded valley to desert. Economic hardships plague the tribe today, but starvation is no longer a threat. Like other reservation Indians in the Southwest, the Pimas eat a cheap but plentiful diet of beans, potatoes, and fried bread. For a people honed to endure scarcity, the result has been disastrous.

Almost all adult Pimas are obese, and the tribe has the highest rate of diabetes ever recorded. Fifty percent of Pimas thirty-five years and older are diabetic, fifteen times the rate in the rest of the United States.

In the late 1970s, the National Institutes of Health launched an intensive long-term study of the tribe to learn more about adult-onset diabetes, especially about two key risk factors involved in the disease: genes and obesity. The results may provide health benefits not only for the Pimas, but also for the millions of other Americans and people around the world who suffer from diabetes. The study may also help geneticists answer another question: Why did evolution leave us with so many "bad" genes?

One possible answer is that many genes that plague us today were "good" when they established a toehold in our blueprints. Modern life does not always reflect the pressures under which man's genetic endowment was formed and refined. The Pimas are a case in point.

Geneticist James V. Neel of the University of Michigan has suggested that diabetes represents a thrifty genetic trait that helped the ancestral Pimas and other peoples to endure periodic famine, allowing them to store food more efficiently and get more mileage from every calorie. Today, with a steady and plentiful food supply, the result of this thriftiness is obesity, disease, and a shortened life-span. An obese Pima can maintain a weight of three hundred to four hundred pounds even with a relatively low calorie intake—a problem that apparently is shared with many other overweight Americans.

Humans are not the only animals who made this genetic adjustment to living with scarcity. The diabetic trait is common among desert animals such as sand rats, who must stuff themselves when food is available, then live off stored fat during the dry seasons. Only in the lab, supplied with food year round, do the sand rats become obese and begin to suffer from the disease humans know as diabetes.

The thrifty-gene theory is also supported by experiments with diabetes-prone mice at the Jackson Laboratory in Bar Harbor, Maine. Douglas Coleman found that on a starvation diet, mice with a double dose of diabetes genes lived eight times longer than normal mice. When Coleman raised the feeding levels to one half the normal lab mouse diet, however, the mice with diabetic genes became obese and diabetic.

Diabetes is not the only "bad" genetic trait that seems to be a mixed blessing, a once-good gene that has outlived its usefulness—at least in some parts of the modern world. Since 1949 a number of investigations have indicated that the sickle cell trait provides carriers with greater resistance against the most severe form of malaria, that caused by the parasite *Plasmodium falciparum*.

Remember, carriers of the recessive sickle cell trait suffer no anemia. They produce both normal and sickle hemoglobin, and only a small proportion of their red blood cells normally distort under low oxygen conditions. Researchers have found that infection by the malaria parasite causes these cells to distort at a much greater rate. When the cells sickle, the parasite inside dies. This does not prevent the disease altogether, since most cells remain normal and the parasite can persist in them, but it lessens the severity of the malaria and perhaps prevents death.

The relative few who inherit two sickle genes and thus develop sickle cell anemia are apparently the price paid for maintaining a large gene pool—about 20 percent of the African population— who are resistant to one of the region's deadliest maladies. In the United States and other countries where malaria is rare, the sickle trait apparently offers its carriers no advantage.

Since other inherited blood disorders such as the thalassemias and G6PD deficiency are also common in parts of the world where malaria is endemic, some scientists have hypothesized that these traits may also play some protective role. Evidence for this, however, is not as strong as with the sickle cell trait.

Are there other defective or variant genes that in small doses provide carriers with certain advantages? The literature is full of speculations. Schizophrenia, for instance: A mild dose of madness is often reputed to be a companion to genius or creativity. Gout, for another: Great men from Isaac Newton to Benjamin Franklin have suffered from it. Some studies have linked elevated uric acid levels in the blood, such as those found in gout sufferers, with high performance and success. Of course, more severe defects involving even higher levels of uric acid, such as those produced in Lesch-Nyhan syndrome, have extreme impacts on the brain and on aggressive behavior that no one would call an advantage.

Only in recent decades have we realized the truth of Sir Archibald Garrod's speculations about each person's genetic uniqueness. Our genomes are flexible, plastic, full of "sports" and variations on a common theme—polymorphisms, as discussed in Chapter 11. Most of them are harmless and neutral—the variant proteins they produce perform as well as the "originals." Occasionally a variation may give an individual living in a certain environment some unique advantage—in processing some type of sensory signal, extracting needed nutrients from food, or defending against a specific strain of parasites. In other cases, as we have seen, a seemingly minor variation can have disastrous effects.

During the 1950s and 1960s, geneticists such as Theodosius Dobzhansky at the Rockefeller Institute proposed that this sort of genetic brinksmanship is the healthiest state for the species. A little variety maintains "hybrid vigor" and gives us a better shot at enduring the vagaries of a changing environment.

But other scientists took a more pessimistic view. Whatever advantage variant genes involved in diseases such as diabetes and sickle cell anemia may have provided at some time in the past, the suffering, debilitation, and health care expense they cause now more than outweigh any benefits they could provide to those who live in developed countries. Among animals and primitive peoples struggling to survive in the wild, these scientists argued, genes that had lost their usefulness or actually harmed their carriers would be expected to decline in frequency over time. But we do not live in the wild, at the mercy of natural selection. Modern medicine and humanitarian social policies shield us from the harsh pressures of the natural environment that initially formed and tempered us.

To some, this fact foreshadows a genetic twilight for the human species, a biological decline that may take us the way of the dinosaurs unless we forestall it. One of the first scientists to express alarm was American geneticist Hermann J. Muller, who began warning of man's genetic deterioration in the early 1930s when eugenics movements were at their height in the United States and Britain. Muller noted that medical advances were allowing victims of diabetes, hemophilia, PKU, and other disorders to survive to marry and bear children, passing on their "bad" genes. And social services were allowing the poor and troubled and down-and-out (a status many of the heralds of genetic twilight equated with genetic weakness) to escape starvation and breed.

". . . Society now comes effectively to the aid of those who for whatever reason, environmental or genetic, are physically, mentally, or morally weaker than the average," Muller reiterated in 1963. "True, this aid does not at present afford these people a really good life, but it does usually succeed in saving them and their children up to and beyond the age of reproduction."

As a remedy he advocated voluntary selective breeding, giving women "germinal choice" by allowing them to be artificially inseminated with genes from men who manifested assorted superior traits, and encouraging more feeble beings not to have children. (Over the decades Muller had some embarrassments over whose sperm should be included in donor banks. In 1935, he suggested Lenin, but withdrew the notion during the Cold War. Later he decided it would be best to freeze donor sperm for a couple of decades until a more objective evaluation of the individual's merits could be made!)

With the dawn of the atomic age, Muller and others—including Julian Huxley, P. B. Medawar, and Linus Pauling—began to worry not only about the genetic defects modern medicine was perpetuating, but also about the new crop of mutations being caused by radioactive fallout. More recently, pollutants, chemicals, and toxic wastes in the environment have evoked more concern about whether our genetic load of potentially harmful mutations is increasing.

Huxley agreed with Muller that the "general quality of the world's population is not very high" and was beginning to deteriorate. "I confidently look forward to a time when eugenic improvement will become one of the major aims of mankind," he said in 1963, admitting, however, that there was no actual evidence for genetic deterioration, only deductive reasoning.

The question of whether we are declining genetically or not remains unanswered and perhaps unanswerable. Few reliable figures are available on the load of deleterious genes carried by modern individuals. And we have nothing from the past to compare them with.

Genetic fitness, as we have seen, is relative. The question is, Fitness for what? There is little doubt that civilization has left many people unfit for the rigorous life of Stone Age hunters. But twentieth-century medicine and social policies can hardly be blamed. Our kind was never satisfied with that life anyway. The moment our ancestors of the genus *Homo* moved into caves and draped themselves in animal skins to ward off the elements, they began engineering their escape from the harsh and random evolutionary processes that created them. We have exercised a good deal of brainpower since then in figuring out ways to make life easier and to compensate for our weaknesses. We invented the wheel, planted crops, and devised dentures, hearing aids, air conditioning, penicillin, and polyester.

". . . In the distant future human beings might require injections and pills for a variety of genetic infirmities—a development that we currently view as unhealthy," University of Washington geneticist Arno G. Motulsky wrote recently. "However, our descendants might consider this state of affairs to be as 'normal' as we consider the wearing of clothing or eyeglasses today."

We don't have to be "fit" for some ancestral world. The environments we inhabit today are largely of our own making, and we can reshape them to maximize whatever genetic endowment the species finds itself with in the future. This doesn't mean, however, that we have to plan for a wheelchair world.

The time it would take to double the frequency of any given genetic disease in the population—assuming everyone who had the disease could be kept healthy long enough to marry and produce an average number of children—varies with the inheritance pattern (recessive, dominant, polygenic, and so forth). But it is always a matter of centuries, if not millennia. If treatments or even cures for such diseases become so sophisticated that victims can bear and raise children, will the increased frequency of the gene then represent any greater genetic deterioration than the prevalence of nearsightedness does today? With a dose of gene therapy available at birth or a periodic infusion of clotting factor cheaply produced by genetically engineered microbes, will hemophilia be a handicap to those who inherit it in the year 2100? These are hypothetical questions, and ones we are not likely to have to face.

Prenatal diagnosis, counseling, and abortion will allow future generations to reduce the burden of severe genetic diseases, and treatments or cures will be available for conditions they cannot or do not choose to prevent. Like the war in infectious diseases, the battle with genetic disorders will never be won. New mutations will arise with every generation, and in a world where children still die of measles and diphtheria, it would be naïve to think that everyone will have access to or make use of prenatal diagnosis or genetic therapies.

Future environments may also turn presently beneficial or neutral genetic traits into diseases—just as an abundant food supply apparently turned the diabetes gene into a liability. In urban America today, genetic resistance to psychological stress, ulcers, and heart disease is more important to fitness than genes for physical strength and quickness or resistance to parasites. The criteria for fitness in the year 2100 will depend on the sort of world we have built by that time.

* * *

Our best insurance that we will have the raw materials to draw on is the characteristic that has sustained us in the past—human diversity. The need for diversity, however, doesn't mean we have to preserve every genetic variation, no matter how hurtful, that random mutation creates. If we are to continue directing our own evolution, we will undoubtedly be more humane and purposeful than the blind forces that initially shaped us.

"Many bad mutations could hardly be accepted in man for the sake of a few good ones, even if we could rely on selection to further the spread of those which are 'good' in our human estimation," Dobzhansky wrote in 1962. "This would mean using human suffering to fertilize the soil in which posterity would grow, an ethically unacceptable procedure. The ethical problem is not, however, acute, since natural selection is no substitute for human knowledge and judgment . . . The biological 'fire' has done wondrously well in the past, making the genetic endowment of mankind adapted to the environments in which men live. But in the future man will have to replace it with a gentler agent."

Two decades later, LeRoy Walters, director of the Center for Bioethics, Kennedy Institute of Ethics at Georgetown University, Washington, D.C., told a congressional hearing: "As for the argument that we lack the wisdom to decide for distant generations, there are some cases in which we can be quite sure about what our descendants in the coming generations will want. The eradication of single-gene defects from the human gene pool seems an eminently worthy goal, at least as worthy as global efforts to eliminate smallpox or measles. If we are concerned that future generations may need the genes that cause single-gene defects in order to survive in novel circumstances, then let us preserve such genes in repositories—as the smallpox virus is preserved—rather than in people."

Diseases were only part of the concern of Muller and other harbingers of genetic twilight. An even more controversial issue involves character, behavior, intelligence, and other human qualities—how much of a role genes play in defining these traits; whether the genetic base for them is deteriorating; and whether there is anything geneticists could or should do about enhancing them. The debate moves us away from the medical uses of genetic engineering to the possibilities for the overall genetic improvement of humanity.

25. Improving the Species?

> ... certainly man is still far from having exhausted all his initial possibilities. But the problem now before us is whether these possibilities themselves could not be enlarged: whether this man, which we are, could not, under his own steam, produce a fresh organic change in himself with all the consequences to which such a change in animal nature might lead in the highest and most specifically human fields.
>
> We shall not concern ourselves here with the question of whether man has possibilities of becoming taller or more robust or more handsome. No; what interests, what fascinates us is the question of whether he can hope to make himself more intelligent, more clever, more sensitive, more disposed to solidarity and altruism—in a word, more human.
>
> —JEAN ROSTAND, French biologist, 1959

> Men would indeed be ignoble if they, Narcissus-like, worshipped their present selves as the acmes of perfection, and reserved their efforts to bring about genetic betterment for their cattle, their corn, and the yeast that gives them beer. Not all men will continue to maintain such smugness of attitude.
>
> —HERMANN J. MULLER, American geneticist, 1959

In an old pump house buried in the backyard of a southern California estate, retired optometrist Robert Graham, who made millions from the invention of shatterproof plastic eyeglasses, began in the late 1970s to implement his vision of human betterment. He called it the Repository for Germinal Choice—a sperm bank to be stocked by the scientific elite, including several Nobel Prize-winning scientists. News of the "Nobel sperm bank" broke in 1980, winning worldwide notoriety for the project.

Graham, then seventy-four, professed at a news conference that his only interest was in "fostering the production of intelligence." He saw regrettably little chance that the United States government would join him in trying to breed a superrace: "We're going completely the other way, to breeding common people of average or lower intelligence."

The reaction of most geneticists to news of the sperm bank was ridicule or amusement. The widow of Hermann J. Muller demanded that Graham abandon his plan to name the sperm bank after her husband. Graham had discussed the project with Muller before the scientist's death in 1967, but Muller had dropped out when Graham disagreed on including "human" traits as well as intelligence among the criteria for sperm donors, Thea Muller told *Science* magazine. The only Nobel Prize-winner who publically admitted donating to Graham's repository was physicist William Shockley, then seventy, who has postulated the genetic inferiority of blacks and works actively for the notion of preventing genetically inferior people from breeding.

The furor over Graham's one-man eugenics movement waned, then was quickly rekindled when he announced the first birth in April 1982. Along with the renewed criticism, however, came a flood of inquiries from women who wanted to be inseminated with "superior" sperm. By the summer of 1983, more than one thousand inquiries had reportedly been received, fifty to sixty inseminations performed, and a half dozen babies born.

Would the sperm of "geniuses," passed out to unscreened women who claimed to be of high intelligence, produce higher-I.Q. children? J.B.S. Haldane noted in 1933, in condemning the notion of a hereditary aristocracy, that genetic shuffling "sees to it that very few human characters breed true." Statistically a man and a woman of high intelligence—by whatever criteria intelligence is defined—do have a greater shot at producing an intelligent child. But the mix of traits passed on from parent to child is purely random. A child might inherit a Nobel Prize-winner's tendency to moodiness or quick temper and not the foundations of his creativity or ambition.

Playwright George Bernard Shaw, a legendary wit, is credited with pointing this out succinctly in an exchange with dancer Isa-

dora Duncan. Duncan reportedly suggested when she and Shaw
met at a dinner party that the two should have a child. With her
looks and Shaw's brains, the dancer bubbled , how could the child
go wrong? "But what if it had my looks and your brains?" Shaw
retorted.

Prospects for improving the species looked slim with the birth
of the repository's first child, a nine-pound girl conceived of an
anonymous "eminent mathematician" and a thirty-nine-year-old
Phoenix woman who had served a jail term for fraud and had lost
custody of two other children for child abuse in her efforts to turn
them into "supersmart kids."

By the summer of 1983, the little repository, which had shuttled
from the pump house to an office building to a beach house in Del
Mar, still had fewer than two dozen sperm donors and no new
Nobelists. It showed no signs of making a major impact on the
human gene pool or American reproductive practices.

The response the sperm bank has generated from potential clients indicates, however, that an age-old human desire to produce
"better" progeny than ordinary matings provide is still strong, no
matter how misdirected, simplistic, and sometimes pathetic the
desire may turn out to be in implementation.

Thinkers since the time of Plato have proposed attempts at improving the human race by selectively pairing "superior" men
and women—positive eugenics—or restricting the breeding of
"inferior" types—negative eugenics. One of the problems with
that proposal, as Haldane and Shaw both noted in their own
ways, is that in the genetic lottery of sexual mating, no one can
predict what combination of traits a child will inherit.

A more troublesome objection is that people through the ages—
or different peoples in any one age—have never agreed on who
was "superior" and what traits were valuable enough to try to
foster. Is intelligence more important than humanitarian qualities,
and how much of the groundwork for these complex traits is
passed along in the genes? ("There is no guarantee that high-I.Q.
people produce better people or a better society," Daniel Callahan, director of the Hastings Institute of Society, Ethics, and the
Life Sciences in Hastings-on-Hudson, New York, said in com-
menting on the so-called Nobel sperm bank. "It is not the retarded kids of the world who produce the wars and destruction.")

Eugenics has a long history of falling into bad hands, becoming
a tool of Nazis, fascists, and racists who have sought to idolize

one particular type of person. Its advocates have championed a range of methods from restricted immigration and selective breeding to sterilization and extermination. The extreme of the eugenic view gives virtually all the credit for intelligence and social and economic potential to one's genetic birthright. It leaves little hope that education and equal opportunity can do much to raise the prospects of the "lower classes," who, according to this view, are lower in the first place because of biology or some supposed "natural order" of things. In the early part of the twentieth century, a strong element among the eugenics movement in the United States and Britain believed that criminality, prostitution, and eroticism were indelibly written in the genes.

A reaction at the other philosophical extreme is the notion of the "blank tablet," touted by diverse visionaries from Marx to educational reformers and cultural anthropologists. In this view, a person's behavior, temperament, intelligence, talents, and accomplishments are shaped, the blank tablet of his mind and character filled in, entirely by education and the cultural influences he encounters after birth. With the right opportunities, almost anyone could become an Einstein, a Rockefeller, or a Mozart.

Few scientists today cling to either extreme view in this ancient nature-versus-nurture debate. As geneticist Theodosius Dobzhansky noted: "Our genes determine our ability to learn a language or languages, but they do not determine just what is said. The structure of neither the vocal cords nor the brain cells would explain the difference between the speeches of Billy Graham and of Julian Huxley."

Arguments about whether genes or culture contribute the most to any single trait are equally futile. The traits we humans are most interested in fostering are, first, only fuzzily defined, and, second, involve complex interplays of unknown genes and largely unknown environmental factors that interact throughout each person's life. Even if we identified all the interacting factors, we couldn't separate, control, or measure them for scientific study. Only in a perfectly equal and uniform environment could we begin to quantify the genetic differences between people. "The closer the approach to equality of opportunity in a society," Dobzhansky wrote, "the more the observed differences between its members are likely to reflect their genetic differences. Inequality of opportunity acts, on the contrary, to hide, distort, and falsify the genetic diversity."

When it comes to promoting some complex character attributes, eugenically or socially, we find that many of the traits we consider desirable seem contradictory. Our natures certainly include some of the requisites for both war and peace, aggression and submission, altruism and selfishness. Scientific or business achievement often involves not only what is measured as "intelligence" on I.Q. tests but also drive, aggressiveness, imagination, competitiveness, or even ruthlessness. People interested in fostering world peace, on the other hand, might wish to select for more altruistic, cooperative, less competitive individuals.

Despite these difficulties, as soon as genetic engineering began to look like a real possibility, visionaries imagined a future world in which human selection and enhancement would be taken far beyond the crude uncertainties of eugenic breeding. They dreamed of creating long-lived, clever, sensitive, and peace-loving descendants. Subhuman types to do our bidding or superhuman species to perform all our specialized intellectual tasks. Miniature people. Space travelers. Ocean dwellers. Polluted-earth residents resistant to radiation, cancer-causing agents, and toxic chemicals. Species in tune with a world of sight, sound, and vibration our sense organs presently shield us from. Others were willing to settle for simply being taller, better-looking, and more desirably configured.

In 1963 Haldane envisioned the possibility of using transferred DNA to "incorporate many valuable capacities of other species without losing those which are specifically human." The animal traits he envied included a finer sense of smell and "the capacity for healing with little scarring which is associated with a loose skin."

He speculated, as we saw in Chapter 4, on modifications that might make humans more suited to extraterrestrial environments—perhaps "gene grafting" to endow man with some of the structural features of a gibbon or a New World monkey for the low gravity of space ships, asteroids, or the moon. For the less likely prospect of reaching and inhabiting a high-gravity world like Jupiter, he suggested dwarfs or quadrupedal men. "The elite will perhaps include anatomical freaks, say people with cerebral hernia whose thinking can be watched with the remote descendant of the microscope, astronauts with prehensile feet unsuited for walking, and so on. . . . I think there will be more psychological polymorphism, and much more tolerance."

Some humans, Haldane thought, would be worth having carbon copies of, if we learn how to clone adults. "Probably a great mathematician, poet, or painter could most usefully spend his life from 55 years on in educating his or her own clonal offspring . . ." He suggested cloning athletes and dancers at an early age and cloning not only accomplished people but those with very rare capacities: " . . . for example, permanent dark adaptation, lack of the pain sense, and special capacities for visceral perception and control." And healthy centenarians. And people resistant to radiation.

Geneticist Sheldon C. Reed of the University of Minnesota enlarged on Haldane's vision in 1968 with his own "far-fetched speculations" that directed mutations could "broaden the range of reaction of a person to many kinds of stimuli," including longer and shorter wavelengths of both light and sound. "Perhaps special sense organs will develop to warn of the presence of radioactive material. Bats have sonar and some fishes emit electrical signals. Perhaps man will evolve some useful form of long-distance communication—a kind of selective and strengthened extra-sensory perception, this taking the place of some of his machines. . . . What unknown personality types have yet to appear as a result of intentional genetic changes?"

Joshua Lederberg and others in the 1960s speculated on the creation of man-animal chimeras or hybrids, perhaps of low-grade intelligence like the semimoron Epsilons of Aldous Huxley's *Brave New World*, to perform society's menial labor.

In 1959 French biologist Jean Rostand noted that the size of the human brain could be doubled simply by causing every cell in the fetal brain to divide one extra time. (Of course, he also suggested at that time that human chromosome sets might be doubled from two to four, a feat which had been performed in plants and had resulted in larger, hardier individuals. It wasn't until a few years later that researchers began to spot the tragic effects of altered chromosome numbers, such as the third copy of chromosome 21 in Down's syndrome.)

The speculations have not toned down much as the science has matured. In 1982 University of Maine biologist Thomas Easton predicted that we will solve the world's population problem in another decade or so by releasing genetically engineered viruses into the water supply or air to make everyone produce miniature offspring. Physicist Freeman Dyson of the Institute for Advanced Studies in Princeton, New Jersey, imagines sitting down at a computer terminal and punching in DNA codes to design new types of humans for planetary exploration. Others foresee implanting "biochips," organic computers, to enhance memory and other brain functions.

If some of these grand abstractions are far-fetched or even repugnant to current sensibilities, they do serve to focus thought on the future uses of a technology that is plunging ahead much more quickly than was imaginable two decades ago. In 1963, listening to Haldane and others predict the biological future, Lederberg said, "It is only by pushing these abstractions to the limit that we are going to be irritated into thinking about questions that are a little bit more general than the immediate ones of today."

In the foreseeable future, any "enhancements" we might want to perform on our species will be sharply limited by our past, by the well-orchestrated groundwork set up by our vertebrate ancestors. We can tinker with evolution, but we don't know how to perform a revolution. Take our basic bone structure, for instance. We won't be able to make a man with wheels simply by synthesizing a gene that alters developmental signals to the point where leg bones circle up (assuming that was simple!). From what little we know of the ground rules of the developmental game, regulatory signals work on sets of things, so we would probably end up circling our arms up, too. We need to learn a great deal more about embryonic development before we can isolate and alter a single event in the complex choreography and attempt to remodel our offspring without achieving grotesque contortions.

Expanding our perception and sensory input sounds more feasible, but the result might be overload and confusion—the kind already seen in drug abusers and suspected in some mental illnesses. We would need to engineer not just our sense organs but the rest of our brain and body systems to process and deal with enlarged inputs. Many of us have enough trouble coping with and deciding how to respond to the stimuli we already receive without having to handle an additional range of sounds, colors, odors, magnetic fields, and so forth.

The broad categories of intelligence and behavior have provoked the most serious near-term speculations by those who would like to improve humanity. We have seen some of the profound impacts that defects in single genes or whole chromosomes can have on intelligence (learning, memory, reasoning, basic comprehension) and behavior (appetite, aggression, motor skills, social skills). The missing enzyme in Lesch-Nyhan syndrome somehow throws its victims into a nightmare world of self-destruction, compelling them to bite off lips and fingers and even to bite and pinch at physicians and family members who come near. Fragile X, Down's syndrome, PKU, and hundreds of other inherited defects cripple the brain and the development of the mind in widely varying degrees. Prader-Willi children are mildly retarded, obese, and eat voraciously and pathologically, devouring sticks of butter, pet food, and edible garbage. Men born with XYY sex chromosomes instead of the normal XY have been found to have a greater-than-average risk of both low intelligence and antisocial (although not necessarily aggressive) behavior. We know there are genes strongly involved in the predisposition to schizophrenia, depression, and even the reading disability dyslexia.

It is clear from what we know of such extremes that genes and biochemistry do affect the way a person thinks, feels, and perceives the world—and as a consequence, the way he behaves in it. But so far we know almost nothing about the roles of genes in normal human behavior.

In 1965 Nobel Prize-winning chemist Max Perutz fantasized a molecular biology examination for students of some future century:

"1. In the murder trial of Rex v Jones, counsel for the defence pleads an Oedipus complex acquired by the defendant in infancy as an extenuating factor. State how you would isolate and identify the complex in molecular form.

"2. Prescribe a therapy, at the molecular level, for (a) Hamlet, (b) Dimitry Karamazov, and (c) Hedda Gabler."

By 1980, Perutz found the questions were only "a few micrometres nearer" to being answered.

The biochemistry and genetics underlying human memory, learning, intelligence, and behavior remain obscure, but scientists are already setting the stage for their exploration. More and more higher-order learning processes that were once thought to be limited to mammalian intelligence are being identified and explored in lowly invertebrates such as the fruit fly, the crayfish, and the sea slug Aplysia. Behavioral neurobiologists are dissecting these learning processes, tracking their biochemical and genetic mechanisms and the environmental signals that trigger them.

Aplysia is actually a shell-less snail that can grow to the size of a human brain. On its back is a respiratory chamber protected by flaps called the mantle. Rimming this "mantle cavity" are the mantle shelf and a built-in siphon that draws in water. The creature usually carries its gills out on its back. However, a light touch to the shelf or siphon causes the snail to jerk its gills back vigorously into the mantle cavity. The reaction is Aplysia's modest version of the defensive escape and withdrawal responses common to vertebrates, and like them, it can be altered by experience and learning.

The learning processes Aplysia shares with higher animals include habituation, where an animal is repeatedly subjected to a weak stimulus and learns to ignore it or respond only minimally; sensitization, in which an animal becomes hypersensitive and responds more vigorously to a harmless stimulus like a touch after being exposed to a potentially dangerous one such as an electrical shock; and classical conditioning, where an animal learns to respond to a stimulus in a novel way by learning to associate it with a more significant event (as Pavlov's dogs learned to salivate at the sound of a bell after they learned to associate the bell with feeding).

When Aplysia is touched in the same location on its mantle shelf or siphon ten to fifteen times with a jet of water, it will become habituated to this stimulus and respond only weakly to repeated squirts. If the animal is then sensitized by an electrical shock to the neck or tail, its responses to the next stimuli, even light touches in formerly habituated sites, will be more vigorous. By repeating the sensitizing shock or increasing its strength, the learning can be made to last—retained in memory—for minutes, hours, even weeks. Scientists can also classically condition the snail's gill withdrawal response by pairing a touch to the siphon with a strong electrical shock to the tail. After fifteen such pairings, the snails will show a greater response to the touch than those who have not learned to associate the touch with a tail shock.

At Columbia University, a team headed by researchers Eric R. Kandel and James H. Schwartz has already identified most of the nerve cells involved in gill withdrawal and is concentrating on the molecular basis of these learning and memory processes. The group's studies indicate these simple forms of learning occur at the junctions or synapses where nerve cells communicate with their neighbors through chemical messengers called neurotransmitters. The memory of what has been learned is retained as changes in the amount of neurotransmitter that is released at specific junctions, changes that alter the strength of existing connections between nerve cells. (Habituation reduces the amount of neurotransmitter that is released and thus reduces the strength of the signal. Sensitization increases it.)

The focus of current research is on deciphering the precise biochemical mechanisms that underlie this prolonged change in neurotransmitter release. Does it require the production of new proteins—and thus the expression of specific new genes—or a modification of existing proteins, or both? Does the conversion of learned experience from short- to long-term memory require an additional molecular event, perhaps the synthesis of a new protein? Will the new protein be produced only temporarily if the memory is not reinforced by subsequent training? Learning, the researchers have suggested, is likely to be a "repertory of mechanisms" instead of a single process.

Evidence so far suggests the biochemistry of these simple learning processes may be the same throughout the animal kingdom and may also play a building-block role in more complex forms of learning and other mental processes. Fruit fly researchers, for instance, have produced a mutant strain called "dunce," with a mutation in a single gene that leaves the creature deficient in sensitization and conditioned learning responses. Unlike normal flies, dunce cannot learn to avoid odors associated with electric shocks. Dunce also lacks one of the biochemicals found to be involved in these learning processes in Aplysia. The biochemical is cyclic AMP phosphodiesterase, an enzyme that controls the breakdown of cyclic AMP, a fundamental regulatory chemical apparently vital to the formation of memories.

Moving from learning to behavior, Richard Axel's group at Columbia, in collaboration with Kandel and Schwartz, has already tracked pieces of the snail's behavioral repertoire down to the genetic level. Aplysia's simple life is taken up with eating and copulating; and when the animal pauses from eating to lay eggs, it goes through a stereotyped ritual that fascinates behavioral geneticists. The hermaphroditic creature, which possesses both male and female sex organs, extrudes a yard-long egg string containing ten million eggs.

As the egg string emerges from the hermaphroditic duct, the snail catches it in its mouth, Axel recounts. At this point the heart and respiratory rate go up. Aplysia waves its head back and forth, continuing to pull the egg string out of the duct and winding it into a tight mass. Then it sticks the entire spaghettilike mass to a solid surface such as a rock. The ritual appears to be made up of seven independent behaviors coordinated in a precisely determined fashion, Axel notes, and it is not modified by experience.

"This is what's known as a classical fixed action pattern," he says. "We want to know what controls this."

Scientists had already learned that the ritual is orchestrated by substances released from a small group of nerve cells. Axel's team isolated, cloned, and determined the molecular code of the gene for one of these substances, the egg-laying hormone, which initiates a single aspect of the behavioral ritual—extrusion of the egg string. (The hormone may also be responsible for raising the heart and respiratory rates.) The gene turned out to carry the blueprint for making not just this hormone but perhaps as many as eleven small proteins or peptides, manufactured as one long chain, then cut apart—a convenient and simple way of regulating the actions of a number of proteins coordinately.

Axel theorizes that most or all of the behaviors involved in egg laying may be accounted for by this constellation of peptides produced by a single gene. "What that implies to me is that you can really build up complicated behavioral repertoires in this manner," by deploying various combinations of peptides, each of which carries messages that stimulate or inhibit various nerve cells or other tissues. One peptide may even function as a part of several different behavioral repertoires, Axel notes, and the possibility is not limited to snails. "If you think about behavior in man, you see that we also combine very similar pieces of our behavior to make up very different behavioral arrays."

Other research groups are pursuing the genetics of behavior in mutant fruit flies, mapping the chromosomal locations of the mutant genes responsible for aberrant behaviors, learning when the genes get switched on in development, what effect each gene product has, and how it causes an abnormal behavior. There is no lack of abnormalities to study. Some 1,500 strains of mutants like dunce, many with bizarre behaviors, are shipped out to geneticists around the world from the fifty-year-old mutant fruit fly repository at Caltech. The mutant strain "drop dead" does just that at the clap of a hand; "shaker" suffers uncontrollable trembles; "stuck" cannot separate after mating; and dumb-but-persistent "buzz off" cannot learn to give up courtship when it has been rejected.

Genes play a crucial role in the behavior and learning processes of higher animals, as any breeder of dogs or fighting bulls knows. Much of that learning is stereotyped learning. Animals are genetically programmed, forced to learn certain vital things at specific times, things too important to be left to chance. They are programmed to learn to recognize the sight or sound of their parents, their young, predators, food sources. However, cultural or environmental elements also play a role in some of these programmed animal behaviors.

In the 1960s Peter Marler of Rockefeller University found that male white-crowned sparrows learn to sing in local dialects, but they are genetically programmed so that they are only able to learn the song of their own species. They must also hear the song at a critical, genetically determined period when they are chicks and then mentally "tape" it or they won't learn to reproduce it later. Dozens of other species of birds have also been discovered to sing in local dialects, and the cues and permissible learning periods for them are also genetically programmed.

Scientists increasingly are probing not just what it is each creature is born to learn, but the environmental cues, neural circuits, biochemical reactions, and eventually the genes involved in these processes.

We already know that genes direct in considerable detail the wiring of the ten billion nerve cells in the human brain. We have genes—perhaps a few crucial ones for each behavior—that ensure that we learn, or pay attention to, or retain certain behaviors or experiences. We also have genes that screen, filter, organize, and limit our perception of the world through our senses.

Our brain, for example, is wired at birth to recognize about forty distinct segments from the continuum of sounds our environment hits us with. All our languages are made up of the sounds that we are programmed to perceive. Specific regions of our brains are wired to process written and spoken language, to decode the meaning of messages we receive, and to make use of grammar and syntax to code meaning into what we want to say. Unlike chimps and other primates, human infants are genetically programmed to learn language without any encouragement.

There is no question that neuroscientists will look for such genes—not for the purpose of tinkering with them but for basic understanding. Some future generation will undoubtedly have a catalog of gene sets and biochemical reactions that underlie each of our behavioral possibilities. At that point, designing a genetic intervention will be quite feasible, if anyone cares to do it. Carrying out such an intervention and having it work as imagined, however, are other matters. Nature is frugal and, as we know, frequently puts one thing to many uses. Any single gene or set of genes we might decide to implant in a healthy person's cells could have more functions than the one we have identified and wish to enhance. The effect is likely to be far less predictable than supplying a normal gene to a patient who lacks it.

"Replacing a faulty part is different from trying to add something to a normally functioning system," W. French Anderson of the National Heart, Lung, and Blood Institute told a congressional committee hearing in 1982. "To insert a gene in hopes of improving or selectively altering a characteristic might endanger the overall metabolic balance of the individual cells as well as of the entire body. . . . It will take a great deal of additional research to understand what the effects will be of altering one or more major pathways in a cell. To correct a faulty gene is probably not going to be dangerous; but intentionally to insert a gene to make more of one product might adversely affect a dozen other pathways."

Behavioral traits may also be hard to tamper with in individuals simply because the wiring pattern of the brain is largely in place by birth. Interventions in this wiring, if we ever choose to try them, would have to wait until we have learned how to transplant or modify multiple genes and we are confident enough in our techniques to try them in human embryos.

"The most important limitation here, I believe, is yet another one, which has not been prominent in discussions of various scary scenarios: motivation," Bernard Davis of Harvard University said at the 1982 congressional hearing. "Every embryo or fertilized egg contains a novel, unknown set of genes, and I find it hard to see why anyone would wish to go to great trouble to insert known behavioral genes into such an unknown background. If a couple wished to have a smarter or stronger child than their own genes seemed likely to provide," it would be simpler to start with someone else's egg or sperm, he suggested. "This kind of molecular manipulation of embryos therefore seems to me extremely unlikely to be tempting, quite apart from its technical difficulties, for the foreseeable future."

Even if science reaches this point of competence, and society finds a compelling reason to try to alter human intelligence, character, behavior, talents, or temperaments, it may still be easier to intervene externally rather than genetically. By the time we have learned that much about the biochemistry of mood and behavior, we will undoubtedly know how to design drugs to make us more sensitive, sociable, clever, kind, and so on. Mood- and behavior-altering drugs, some therapeutic and some illegal, are available now and will undoubtedly be improved in the future. Education, behavior modification, indoctrination, and psychosurgery have also proved to be powerful tools for modifying whatever we define as "human nature." Many scientists working in the forefront of behavioral genetics believe there will always be more efficient ways of reshaping human traits, if we care to, than tinkering with genes.

"If you think about the brain, culture has a far more profound effect on future generations, I think, than does the genetic constitution," Axel says. "Our cultural information obviously is far greater than the information of our chromosomes. We can't put into our genomes all the information we've put into our libraries. We can't even put into our heads all the information we've put into our libraries. Cultural transmission of information is profound. If we want to deal with changing people, we ought to be dealing at the level of culture and not at the level of genes, and that is possible right now. You'd have to do a lot of embryos to match the impact some of the religious cults have had."

A few scientists do think that changing the behavioral foundations evolution has given us is important enough to survival of the species that we must use some drastic means to accomplish it, if not by genetic engineering, then by cultural or value engineering. Sociobiologist Edward O. Wilson is one of these few.

"It's extremely likely that within ten years, twenty at the outside, a number of genes will have been identified whose effects can be traced through the actual production of particular chemicals in the brain and thence to measurable properties of temperament, mood and even cognitive ability," Wilson said in 1982.

"If we could change our basic nature—the strength of the sex bond, the pleasure you get from children—through genetic intervention, then with more knowledge of the genetic basis of the assembly of the mind, you could come up with human beings who respond to the world in very different ways—some taking deep pleasure in living in a city, for example, others who are able to live in rural communes. Suppose we really could do it. Should we? And if so, what direction should we choose?" If we don't make changes somehow, he believes, our Stone Age behavioral legacy will continue to threaten our future.

Despite scientific hurdles that make attempts to remodel man unlikely in the foreseeable future, philosophers, bioethicists, theologians, futurists, and others have been concerned with the ethics of such a prospect for the past two decades. The possibility of "improving" our species obviously raises more troubling ethical questions than using the new knowledge to cure diseases that cause suffering and degradation in individuals.

Who will get these "improvements"? Only those who can pay? Most of the basic research that might make such intervention possible is being done at taxpayer expense. Perhaps enhancements should be reserved for those who "deserve" them most, by whatever criteria of merit we choose—intelligence, creativity, or talent. Genetic enhancement might thus make the deserving more so, thus reinforcing native differences and almost certainly as a result, economic, social, and cultural inequalities among people. Perhaps, following up on the commitment in our society to provide equality of opportunity despite a person's social or economic disadvantages at birth, we might choose to use our new knowledge to level the impacts of genetics, too.

What improvements will have priority? Various individuals and cultures would have different answers, based largely on their values and their perceptions of what the world and human beings should be like in the future. Philosopher Stephen Stich of the University of Maryland, College Park, speculates that several generations of divergent choices could bring on "the genetic fragmentation of the human species" so that members of different cultural groups "will no longer be interfertile [able to mate and produce offspring]."

If advertising is a good indication of what Western culture worships, our values run to youth, beauty, strength, and financial success. We already spend billions on vitamins, spas, cosmetics, plastic surgery, sporting goods, and self-improvement courses trying to achieve these ends nongenetically. Are those the values we will want to foster genetically if it becomes possible to do so? If "everyone is doing it," would a person be truly free to reject genetic enhancement for him or herself or the children?

Stich foresees the development of intense social pressures to use our new genetic powers as the technology grows in sophistication in the coming decades. "During the last year or two we have seen an explosion of interest in home microcomputers; many of the people who buy these wonderful, expensive machines do so in the hope that they will give their children a competitive edge in a technologically competitive world. Closer to the fringes of our society we have seen that some women are prepared to have themselves impregnated with the sperm of a Nobel Prize winner in the hope of bearing an intellectually gifted child.

"Both of these phenomena underscore the fact that the desire to help one's children to excel is a powerful and widespread motivational force in our society. When, via genetic engineering, we learn how to increase intelligence, memory, longevity, or other traits conveying a competitive advantage, it is clear that there will be no shortage of customers ready to take their place in line. Moreover, those who are unwilling or unable to take advantage of the new technology may find that their offspring have been condemned to a sort of second class citizenship in a world where what had been within the range of the normal gradually slips into the domain of the subnormal."

Will we be tempted, if it ever becomes possible, to use genetic tinkering as a panacea for social problems? Eliminate racism by leveling racial or ethnic distinctions? Eliminate the need to clean up the workplace and the world by making people pollution-resistant? Eliminate strife by making people less aggressive and more cooperative? Would we be willing to give up a great deal of humanity, just as the characters in *Brave New World* did, for the sake of stability, peace, and freedom from want or pain?

If we should ever learn to tamper with the basic life processes of development, growth, and aging, or with the elements individuals most closely link with self—personality or intelligence—we would have to change our fundamental concepts of personal identity and the nature of our relationships with each other. "The current tendency is to think of a person as an individual of a certain character and personality that, following the normal stages of physical, social, and psychological development, is relatively fixed within certain parameters," the President's Commission for the Study of Ethical Problems in Medicine and Biomedical and Behavioral Research reported in 1982. "But this concept—and the sense of predictability and stability in interpersonal relations that it confers—could quickly become outmoded if people use gene splicing to make basic changes in themselves over the course of a lifetime."

Would the availability of genetic enhancement of specific traits change our ideas of personal freedom and individuality? If an embryo were given an extra dose of intelligence, musical talent, or hand-eye coordination, would that mean its choices of career, education, etc., had been taken away? Would a child be free to choose to be a poet rather than a basketball player if society had invested extra growth hormone and coordination genes in him or her? If so, would such enhancement be cost-effective or wasted? Would we come to think of human beings in narrow, functional terms instead of as complex individuals with broad potentials?

Alexander Capron, executive director of the 1982 presidential commission reporting on human uses of genetic engineering, commented that designing changes for future generations may be "like shooting at a moving target. If we think greater height is desirable, because people can then go out and be basketball players and make a lot of money, in fifty years it may turn out that it's a terrible thing to have all these tall people around because we really ought to fit inside of spaceships and get shot up to the moon to get away from the pollution. The very fact that one has to make light of it just shows that you can't begin to predict the future that accurately."

Despite this troublesome array of questions future generations may be forced to deal with, few thoughtful people are willing to hold victims of genetic disease hostage to these fears by putting a halt to the development of gene therapy.

26. Choosing Our Future

Man and man alone knows that the world evolves and that he evolves with it. By changing what he knows about the world man changes the world that he knows; and by changing the world in which he lives man changes himself. Changes may be deteriorations or improvements; the hope lies in the possibility that changes resulting from knowledge may also be directed by knowledge. Evolution need no longer be a destiny imposed from without; it may conceivably be controlled by man, in accordance with his wisdom and his values.

—THEODOSIUS DOBZHANSKY, American geneticist, 1962

Imagination may be our saving grace in a world of rapid change. Our ability to fantasize uses for a new technology long before we can think about implementing them gives us time to shudder, moralize, joke, consider, and react with some consensus to the implications.

For a quarter century the thinkers among us have been speculating about what we might do as soon as we had the power to manipulate human genes. Now we are on the threshold of doing it. We won't make Hitler clones, Wonder Women, man-ape hybrids, or any of the other provocative oddities of headline and movie fame anytime soon. We will use the technology to prevent or cure human disease. We have plenty of time left to decide what else we would like to do with our new powers, if anything, as understanding and manipulative skills improve.

Our society has come to no rigid conclusions about what must or must never be done with the technology, and that is as it should be. We are changing the world, and our needs and visions will change with the nature of the world we create. What we have done over the past twenty-five years is develop a rather thorough listing of our anxieties and the questions we want answered before trying to make elaborate forays into human manipulation.

Perhaps our greatest anxiety about this new technology is that it will have a personal impact on all of us, forcing us to shoulder another responsibility. Our genes are out of sight and have been largely out of mind. Now we will be burdened with thinking about and making genetic decisions we once left to fate or deity. Many of us have felt bitterly disappointed, cheated, or punished at the genetic "decisions" imposed by those entities: children born under a sentence of retardation, death, or suffering; a heart attack in the flush of youth; even such minor miseries as a plain face or a too short body.

But there has been some resigned comfort in the fact that the genetic roulette was out of our hands. All we were expected to do was to maximize whatever potential we inherited. Increasingly, however, we will be called on to make choices about our genetic fate and that of our children. We will have to make personal decisions about what it means to be human, what should be inviolable about ourselves, and which of our differences, weaknesses, or vulnerabilities are important to our existence as unique individuals.

Our anxiety is heightened by the fact that genetic engineering is not arriving on the scene alone. It is part of a cascade of scientific and social developments that have engulfed our society during the past two decades, revolutionizing the reproductive process and changing the nature of family ties: contraception, artificial insemination, abortion, sex determination, prenatal diagnosis, test-tube babies, surrogate mothers, frozen embryos.

The high technology of birth and death—intensive-care units stocked with machines that can sustain critical bodily functions without our cooperation—has changed the natural course and even the definitions of when life begins and ends. A two-pound fetus born three months early may live to become a healthy individual. A person with a heartbeat but no brain function may be declared dead.

We have had to learn to embrace new technologies that leap from science fiction fantasies to the clinic or bedroom with only a few years of reaction time. Barely a decade ago, test-tube babies were the subject of much wide-eyed speculation, but the possibility seemed too remote for most of us to worry about. Then news of Louise Brown's birth hit the front pages in 1978. Ethicists, theologians, and philosophers met to ponder and discuss, fundamentalist preachers condemned, and childless couples throughout the Western world lined up to request this new "in vitro fertilization" service. Babies conceived in a lab dish are still uncommon, numbering in the dozens worldwide. But they are already old news. Their birth announcements make headlines only when some new twist is added—test-tube twins, babies born from embryos frozen before implantation in the mother's womb, or embryos transferred from one woman to another.

Gene therapy is also approaching much more quickly than anyone imagined. "Human genetic engineering was once thought to be impossible, the product of science fiction writers' vivid imaginations," Representative Albert Gore, Jr., told the House science investigations subcommittee as he opened a hearing on human uses of genetic engineering in 1982. "Like so many other accomplishments in science and technology, however, human genetic engineering has moved from the realm of impossibility to certainty. The real question today thus is not whether human engineering will become a reality but, rather, when it will be so and for what purposes it will be used."

In July 1980, leaders of the American Jewish, Catholic, and Protestant church organizations—the National Council of Churches, the Synagogue Council of America, and the United States Catholic Conference—wrote a letter of concern to then-President Carter about the rapid growth of genetic engineering and the uses to which it might be applied. "History has shown us that there will always be those who believe it appropriate to 'correct' our mental and social structures by genetic means, so as to fit their vision of humanity," the letter stated. "This becomes more dangerous when the basic tools to do so are finally at hand. Those who would play God will be tempted as never before." The groups called for a thorough study of the issues to determine what oversight and controls might be needed.

Carter assigned the task to the President's Commission for the Study of Ethical Problems in Medicine and Biomedical and Behavioral Research. In November 1982, the commission reported its findings to President Reagan and to Gore's subcommittee.

"Some people have suggested that developing the capability to splice human genes opens a Pandora's box, releasing mischief and harm far greater than the benefits for biomedical science," the report noted. "The Commission has not found this to be the case."

The commission also "could find no ground for concluding that any current or planned forms of genetic engineering, whether using human or nonhuman material, are intrinsically wrong or ir-religious per se. The Commission does not see in the rapid devel-opment of gene splicing the 'fundamental danger' to world safety or to human values that concerned the leaders of the three reli-gious organizations.

"It is true that genetic engineering techniques are not only a powerful new tool for manipulating nature—including means of curing human illness—but also a challenge to some deeply held feelings about the meaning of being human and of family lineage," the report noted. "But as a product of human investigation and ingenuity, the new knowledge is a celebration of human creativ-ity, and the new powers are a reminder of human obligations to act responsibly."

The report concluded that the medical uses of genetic engineer-ing now being planned "resemble accepted forms of diagnosis and treatment" and should be judged by the same ethical standards and safeguards. It added that some hypothetical future uses of the technology "such as treatments that can lead to heritable changes in human beings or those intended to enhance human abilities rather than simply correct deficiencies caused by well-defined ge-netic disorders" deserve close scrutiny.

"Gene splicing is a revolutionary scientific technique that re-casts past ideas and reshapes future directions. Even so, it does not necessarily follow that all its applications or objectives repre-sent a radical departure from the past. . . . A complex and seem-ingly mysterious new technology with untapped potential is a ready target for simplistic slogans that try to capture vague fears. This is very much the case with genetic engineering."

The commission explored again a variety of concerns, from the "concrete and practical" to the "vague and imprecise," that have been expressed over the past few decades about potential human uses of genetic engineering. Most of the concerns anticipate a time when humans might be tempted to take the technology be-yond its obvious medical applications.

"Playing God" was a phrase used by the three religious groups, and is often seen in popular articles and books on genetic en-gineering. The commission asked the groups to delve further into the nebulous phrase based on each of their theological traditions and define their specific concerns. In the end the theologians as-signed to the task found no reason to condemn advances in molec-ular biology as another Promethean incident of man stealing power from the gods.

"In the view of the theologians, contemporary developments in molecular biology raise issues of responsibility rather than being matters to be prohibited because they usurp powers that human beings should not possess," the report noted. None of the major Western religions puts taboos on the use of man's intelligence to delve into the mysteries of nature. Such a quest for knowledge is in fact encouraged, along with responsible use of that knowledge.

The "playing God" phrase seemed to the commission to encompass a wide range of religious and secular concerns rather than any specific theological issue. "At its heart, the term represents a reaction to the realization that human beings are on the threshold of understanding how the fundamental machinery of life works," the report said. "A full understanding of what are now great mysteries, and the powers inherent in that understanding, would be so awesome as to justify the description 'God-like.' In this view, playing God is not actually an objection to the research but an expression of a sense of awe—and concern. . . . By identifying DNA and learning how to manipulate it, science seems to have reduced people to a set of malleable molecules that can be interchanged with those of species that people regard as inferior. Yet unlike the earlier revolutionary discoveries [Copernicus, Darwin], those in molecular biology are not merely descriptions; they give scientists vast powers for action."

June Goodfield expressed the same idea in her book entitled *Playing God*: "Earlier revolutions in science, such as the Copernican, changed our perspective of ourselves in the world, or, as with the Darwinian, changed the knowledge of our origins, but this new revolution may give us the possibility to change ourselves, to alter our nature as we will, in some consciously directed form."

Molecular biologist Robert Sinsheimer, chancellor of the University of California at Santa Cruz, was once an advocate of genetic manipulation, but now stands as one of the severest critics of the field. Sinsheimer believes that genetic intervention may dehumanize man and cause us to lose our reverence for life. "In the hands of the genetic engineers, life forms could become extraordinary Tinkertoys and life itself just another design problem," he wrote in 1983.

If "playing God" conveys anxiety about creating new life forms, the presidential commission noted that plant and animal breeders and hybridizers have been doing this by slower methods for centuries. It could find no rational or religious grounds to conclude that there is anything "intrinsically wrong with intentionally crossing species lines . . . unless one is willing to condemn the production of tangelos by hybridizing tangerines and grapefruits or the production of mules by the mating of asses with horses."

The far-off possibility of producing human-nonhuman hybrids, however, does have special moral dimensions. "Could genetic engineering be used to develop a group of virtual slaves—partly human, partly lower animal—to do people's bidding?" the report asked. "Paradoxically, the very characteristics that would make such creatures more valuable than any existing animals (that is, their heightened cognitive powers and sensibilities) would also make the moral propriety of their subservient role more problematic. Dispassionate appraisal of the long history of gratuitous destruction and suffering that humanity has visited upon the other inhabitants of the earth indicates that such concerns should not be dismissed as fanciful."

Implanting a human interferon gene into a bacterium or into a mouse embryo, or fusing mouse and human cells into hybridomas—the limits of our powers right now—hardly seem to qualify these creations as being part human in the sense that causes people anxiety. And as the commission noted, we do not yet have a good idea of what, if any, traits are "uniquely human, setting humanity apart from all other species." As we saw in Chapter 25, many traits once regarded as strictly mammalian are being found in so-called lower animals, from flies to snails.

The report also noted that some people are concerned that scientists, like the legendary Dr. Frankenstein, do not have enough knowledge to predict fully and control the consequences of whatever new life they might create. Others fear deliberate Nazi-style abuses of the technology rather than accidental mishaps. But it is likely that dictators and cults will continue to find it quicker and more certain to use mental manipulation, drugs, fear, or repression to accomplish their ends. As Leon R. Kass of the University of Chicago noted in 1971, "Biomedical technologies would add very little to our highly developed arsenal for mischief, destruction, and stultification."

A more realistic but subtle danger if genetic manipulation should become relatively easy, the commission noted, is that "there may be a tendency to identify what are in fact social problems as genetic deficiencies of individuals or to assume that the appropriate solution to a given problem, whether social or individual, is genetic manipulation."

Beyond the "playing God" objection, some people express reservations about tampering with four billion years of "evolutionary wisdom." Evolution, however, is not a directed process aimed at any particular goal. As we saw in Chapter 24, traits that make for fitness in one age or place are often detrimental in other ages and environments. Since we have assumed increasing control over the environments we live in, it makes sense that we be able to influence the genetic side of the equation as well. And as we have also seen, we humans are long past relying on random mutations and natural selection to meet the challenges of survival.

Of more immediate and practical concern should be the social and political issues raised by the medical potential of genetic engineering. In Chapter 11 we saw that genetics is already opening up opportunities for the prediction and prevention of disease that our current health-care system is not structured to use to full advantage. Gene therapy will call for another shift in orientation and resources. At first, it will be an expensive novelty like the artificial heart, and we will practice a sort of battlefield ethic of triage in deciding which few patients should receive it. But many geneticists believe gene therapy will eventually become a fundamental part of medical services, as profound in its impacts as polio vaccines and penicillin. Will everyone have equal access to the benefits of this therapy?

"If the promise of gene therapy is realized, a revolution in medicine parallel to that effected by antibiotics and vaccines may lie ahead," LeRoy Walters of Georgetown University, Washington, D.C., has predicted. "If gene therapy is successful, it will be no half-way technology like the iron lung or the dialysis machine but rather a cure, at the most fundamental level, for certain genetic diseases. As such, it will, in my view, become a part of the basic health care which our society owes to the genetically afflicted.

"In the early stages of its diffusion, gene therapy may be quite expensive and may in fact be available only to the upper tier in the two-tier health-care system . . . If we as a society are committed to the notion of equal access to basic health care, then we should perhaps begin now to envision ways of making gene therapy available to all who suffer from genetic disease."

Others worry about an overallocation of resources to genetic diseases, urging that these afflictions be viewed in the perspective and context of all human disease and suffering. Genetic diseases affect approximately 12 percent of the world's population, not a large proportion compared to the number affected by infectious diseases, parasites, and starvation. Of the generation being born now—the generation that will reach adulthood after the year 2000—90 percent will grow up in the less-developed nations of the earth.

Jon Gordon of the Mount Sinai School of Medicine believes the emphasis should be on prenatal screening and abortion, not the more expensive service of gene therapy for individuals or embryos. He worries that the lure of a new technology will distract attention and money from less glamorous but more pressing medical and social problems.

"Do we want to do all this because it's pleasing to us to think about the power of the human mind and the collective human enterprise? Or are we trying to improve the lives of human beings? I still believe this is the idea we have to keep at the forefront. If we don't, we're going to slip into the attractive and imaginative ways of thinking that a lot of times allow us to ignore some of the really difficult problems in human populations and society, problems that are of urgent importance to solve. We have the technology now so that no one would be hungry on earth. Millions of people suffer from diseases that wouldn't even be seen in an advanced society—cholera, malaria.

"People have used every escape mechanism imaginable not to face problems like these that have a solution . . . ," Gordon says. Just because solutions to genetic disease are developed, it doesn't mean the problem will be solved. "You have to organize society in such a way that solutions are actually applied to problems."

In order to keep such "social and ethical implications of gene splicing before the public and policymakers as these developments become feasible in the years ahead," the President's commission recommended creation of an oversight group that would have education as a primary responsibility.

In 1983 under Gore's sponsorship, Congress took the first steps toward setting up an independent advisory group dominated by nonscientists. This group would monitor developments and keep a continuing dialogue going on potential human applications of gene splicing. Interest in establishing such a commission was boosted when leaders of most of the major church groups in the United States, stirred to anxiety by activist Jeremy Rifkin, signed a resolution calling on Congress to ban any genetic engineering of human sperm or egg cells, even to correct inherited disease.

In many ways our increasing control over nature, from the fissioning of atoms to the quality of the air we breathe, has left some individuals feeling less in control. Only a few of us understand any of these developments well enough to try to exert deliberate influence over their use. The rest of us are forced to rely on these experts, and we feel dragged along into a future we're not sure we understand or want. For all the hoopla about "playing God" or playing Frankenstein, our real fears of genetic engineering seem to be about the domination of technology in our lives and the loss of personal control. They are the same ambiguous fears we have when we marvel at but recoil from the artificial heart and other heroic technologies.

Our genetic knowledge won't disappear. The only way we can partake of the power it confers and channel it into personal uses that suit our visions and values is to understand it. If we rebel against genetic screening and therapy as unnatural and resent the intrusion of technology in our lives, then we must be prepared to accept the "natural" conditions of humanity: occasional child-lessness, disease, suffering, and untimely death. Could an infusion of new genes possibly be more dehumanizing than an abbreviated life of severe retardation or unrelieved suffering?

"I do not subscribe to the view that a technological society is progressively dehumanizing us," bioethicist John Fletcher of the National Institutes of Health said in 1972. "Man makes culture and has the initiative in the culture-making process; the roots of dehumanization lie deeper in us than feedback from technique or technology. There is no ultimate threat to the concept of the sanctity of human life or of the individual from genetic knowledge. Man is an ultimate threat to himself when he is unwilling to accept the anxiety of living in the constantly shifting territory between learning to use his freedom and learning to respect his limits. The anxiety we feel, when we approach such problems bearing such meager wisdom, is a signal that we are in touch with the basic human condition. And this will not change."

Genetic engineering will not allow us to escape our biological fate entirely, and most of us wouldn't want to. In applying advances in genetics, we will have to use our best judgment to strike a balance between fatalism, accepting the hand dealt us in the genetic shuffle, and an unrealistic search for some sort of engineered perfection. It will be up to each of us to make sure that the genetic options we choose for ourselves and our children serve our humanity, uphold our individual values, and fit into the goals we set for our lives.

Source Notes

PAGE *Chapter 1* **Gene Therapy: Expanding Our Options**
11   "Man can create . . ." Theodosius Dobzhansky, *Mankind Evolving* (New Haven and London: Yale University Press, 1962), 338.

Chapter 2 **The Frontier Within**
14   "It does not . . ." Edward L. Tatum, "A Case History in Biological Research," *Science* 129 (26 June 1959): 1711–14.
15   "Infectious diseases, altogether . . ." For a detailed look at our continuing war on microbes, see Allan Chase, *Magic Shots* (New York: William Morrow & Co., 1982).

Chapter 3 **Unraveling Inheritance**
17   "Our ancestors understood . . ." L. P. Coonen detailed the earliest roots of genetics in "Protogenetics from Adam to Athens," *The Scientific Monthly* 83 (August 1956): 57–65.

Chapter 4 **Designer Genes**
23   "Perhaps within the . . ." Tatum, "A Case History," 1714.
23   "The late British . . ." J.B.S. Haldane's speculations can be found in "Biological Possibilities for the Human Species in the Next Ten Thousand Years," in *Man and His Future*, Gordon Wolstenholme, ed. (Boston: Little, Brown & Co., 1963), 337–61. University of Maine biologist Thomas Easton is quoted in Eric Mishara, "Small People," *Omni* (January 1983): 92. Andy Rooney made his genetic wishes in a column released by the Chicago Tribune-New York News Syndicate, Inc., in December 1982.
26   "Manipulations in humans . . ." President's Commission for the Study of Ethical Problems in Medicine and Biomedical and Behavioral Research, *Splicing Life, A Report on the Social and Ethical Issues of Genetic Engineering with Human Beings* (Washington: GPO, November 1982), 2.

Chapter 5 **The Genetic Lottery**
29   ". . . if we assign . . ." *The Works of Aristotle Translated into English*, trans. A. Platt (Oxford: Clarendon, 1910), quoted in Coonen, "Protogenetics," 64.

PAGE
29   "A mother in . . ." The quotes from Archibald E. Garrod and the account of his work found in this chapter are taken from his paper on "The Incidence of Alkaptonuria: A Study in Chemical Individuality," *The Lancet* 2 (13 December 1902): 1616–20.

34   "In 1958 when . . ." Tatum, "A Case History," 1712.

Chapter 6   Inborn Errors

36   ". . . the idea occurred . . ." Linus Pauling, "Fifty Years of Progress in Structural Chemistry and Molecular Biology," *Daedalus* (Fall 1970): 1011.

38   "At Johns Hopkins . . ." Victor A. McKusick, *Mendelian Inheritance in Man*, 6th ed. (Baltimore: Johns Hopkins University Press, 1983); McKusick, telephone interview, April 11, 1983.

39   "—Four to 6 percent . . ." Unless otherwise referenced, the statistics on incidence and costs of genetic defects used in this chapter have been compiled from McKusick, *Mendelian Inheritance*; National Institute of Child Health and Human Development, *Antenatal Diagnosis, Report of a Consensus Development Conference*, NIH Publication No. 80-1973 (Washington: GPO, 1979), I-1 to I-263; Elizabeth M. Short, "Genetic Disorders: The High Risk Couple," in *Medical Complications During Pregnancy*, 2nd ed., Gerard N. Burrow and Thomas F. Ferris, eds. (Philadelphia: W. B. Saunders Co., 1982), 115; National Research Council Committee on Chemical Environmental Mutagens, *Identifying and Estimating the Genetic Impact of Chemical Mutagens* (Washington: National Academy Press, 1983), 1982–83.

41   "Nearly 225 of . . ." Victor A. McKusick, "The Last Twenty Years: An Overview of Advances in Medical Genetics," in *Progress in Clinical and Biological Research* 45, E. S. Russell, ed. (New York: A. R. Liss, 1981), 127–44; McKusick, telephone interview, April 11, 1983.

41   "When receptors for . . ." Michael S. Brown and Joseph L. Goldstein, "Familial Hypercholesterolemia: A Genetic Defect in the Low-Density Lipoprotein Receptor," *New England Journal of Medicine* 294 (17 June 1976): 1386–90.

44   "In 1976, a Danish . . ." Herman A. Witkin et al., "Criminality in XYY and XXY Men," *Science* 193 (13 August 1976): 547–55.

45   "Twins' physical similarities . . ." Constance Holden, "Minnesota Twin Study," *Science* 207 (21 March 1980): 1323.

45   "If one identical . . ." William L. Nyhan with Edward Edelson, *The Heredity Factor* (New York: Grosset & Dunlap, 1976), 189–94.

45   "Researchers estimate 20 . . ." Ibid., 202.

PAGE   *Chapter 7* **Inborn Vulnerabilities**
48   "Only a few . . ." Joshua Lederberg, "Participatory Evolution; What Controls for Genetic Engineering?," *Current* 121 (September 1970): 49.
48   "From what scientists . . ." The section on HLA is drawn from many sources, including: Arne Svejgaard, Per Platz, and Lars P. Ryder, "HLA and Disease 1982—A Survey," *Immunological Reviews* 70 (1983): 193–218; Donna D. Kostyu and D. Bernard Amos, "The Histocompatibility Complex," in *The Metabolic Basis of Inherited Disease*, 5th ed., John B. Stanbury et al., eds. (New York: McGraw-Hill, 1983), 77–95; Jean L. Marx, "Cloning the Genes of the MHC," *Science* 216 (23 April 1982): 400–402; Andrea James, "New Era for Preventive Genetics," *Stanford MD* (Spring/Summer 1979): 20–23; William J. Cromie, "Genetic Trade-Offs: The Hidden Cost of Survival," *Sciquest* (January 1981): 10–14; Zsolt Harsanyi and Richard Hutton, *Genetic Prophecy: Beyond the Double Helix* (New York: Rawson, Wade, 1981), 55–78.
51   "Carl Grumet at . . ." Carl Grumet, telephone interview, April 27, 1983.
52   "Finally, some researchers . . ." Richard Conniff, "Supergene," *Science Digest* (March 1982): 111; Kathleen Stein, "Supergene," *Omni* (December 1980): 81.
53   "Genes, viruses, and . . ." The material on diabetes is drawn from J. Kobberling and R. Tattersall, eds., *The Genetics of Diabetes Mellitus* (New York: Academic Press, 1982); Lester B. Salans, "Diabetes Mellitus: A Disease That Is Coming into Focus," *JAMA* 247 (5 February 1982): 590–94; Howard J. Sanders, "Diabetes: Rapid Advances, Lingering Mysteries," *Chemical & Engineering News* (2 March 1981): 30–45; Jeffrey S. Flier, "Looking Anew at Insulin Resistance," *Patient Care* (15 March 1982): 149–197.
54   "George S. Eisenbarth and . . ." S. Srikanta et al., "Islet-Cell Antibodies and Beta-Cell Function in Monozygotic Triplets and Twins Initially Discordant for Type I Diabetes Mellitus," *New England Journal of Medicine* 308 (10 February 1983): 322–25.
56   "Various groups of . . ." Peter S. Rotwein et al., "Polymorphism in the 5' Flanking Region of the Human Insulin Gene: A Genetic Marker for Non-Insulin-Dependent Diabetes," *New England Journal of Medicine* 308 (13 January 1983): 65–71.
56   "The impact of . . ." Sources for the section on heart disease include: Arthur G. Steinberg et al., eds., *Genetics of Cardiovascular Disease*, vol. 5, Progress in Medical Genetics, new series (Philadelphia: W. B. Saunders Co., 1983); Joseph L. Goldstein

and Michael S. Brown, "Genetics and Cardiovascular Disease," in *Heart Disease, A Textbook of Cardiovascular Medicine*, vol. 2, Eugene Braunwald, ed. (Philadelphia: W. B. Saunders Co., 1980), 1683–1722; W. Jape Taylor, "Genetics and the Cardiovascular System," in *The Heart, Arteries and Veins*, 4th ed., J. Willis Hurst et al., eds. (New York: McGraw-Hill, 1978), 753–75.

56 "First, coronary heart . . ." The statistics on incidence of heart disease in this section are from the American Heart Association, *Heart Facts 1983*.

57 "(A study reported . . .") "Strokes in Kids Tied to Familial Lipid Anomalies," *Medical World News* (18 January 1982): 102–103.

59 "For centuries people . . ." Material in this section is based on Paul H. Wender and Donald F. Klein, *Mind, Mood, and Medicine* (New York: Farrar, Straus, Giroux, 1981), 173–91; Seymour S. Kety, "Disorders of the Human Brain," *Scientific American* 241 (September 1979): 202–14; Kety et al., eds., *Genetics of Neurological and Psychiatric Disorders* (New York: Raven Press, 1983).

62 "In 1981 researchers . . ." Lil Traskman et al., "Monoamine Metabolites in CSF and Suicidal Behavior," *Archives of General Psychiatry* 38 (June 1981): 631–36.

63 "One of several . . ." Wray Herbert, "Test for Depression Called Unreliable," *Science News* 123 (21 May 1983): 326.

63 "In 1981 Lowell . . ." Lowell R. Weitkamp et al., "Depressive Disorders and HLA: A Gene on Chromosome 6 That Can Affect Behavior," *New England Journal of Medicine* 305 (26 November 1981): 1301–1306.

64 "Studies in the . . ." Marc A. Schuckit, telephone interview, April 28, 1983; Schuckit, "Alcoholism and Genetics: Possible Biological Mediators," *Biological Psychiatry* 15 (June 1980): 437–47.

64 "Schuckit has found . . ." Marc A. Schuckit and Vidamantas Rayses, "Ethanol Ingestion: Differences in Blood Acetaldehyde Concentrations in Relatives of Alcoholics and Controls," *Science* 203 (5 January 1979): 54–55.

Chapter 8 **Genes and Cancer**

66 "You may feel . . ." Albert Szent-Gyorgyi, "The Promise of Medical Science," in *Man and His Future*.

66 "In 1981 they found . . ." The material on cancer genes was compiled largely from presentations by David Baltimore, Mariano Barbacid, Carlo M. Croce, Robert C. Gallo, Denise A. Galloway, Mary E. Harper, A. Harry Rubin, Alan Y. Sakaguchi, George J. Todaro, Harold E. Varmus, and George F. Vande Woude, Jr., at the American Cancer Society's 25th Science Writers' Seminar in San Diego, March 20–23, 1982.

PAGE
72   "Douglas Lowy's team : . . ." Jean L. Marx, "Change in Cancer
        Gene Pinpointed," *Science* 218 (12 November 1982): 667.

72   "In 1970, before . . ." Jorge J. Yunis, "The Chromosomal Basis
        of Human Neoplasia," *Science* 221 (15 July 1983): 227–236.

73   "Jorge Yunis of . . ." "Cancer Genes," *Science Digest* (July
        1981): 27.

74   "Croce's hypothesis about . . ." Jean L. Marx, "The Case of the
        Misplaced Gene," *Science* 218 (3 December 1982): 983–85.

74   "In 1983 a group . . ." Russell F. Doolittle et al., "Simian Sar-
        coma Virus *onc* Gene, v-*sis*, Is Derived from the Gene (or Genes)
        Encoding a Platelet-Derived Growth Factor," *Science* 221 (15
        July 1983): 275–76; Jean L. Marx, "*Onc* Gene Related to Growth
        Factor Gene," ibid., 248.

75   "Experiments reported by . . ." Harmut Land, Luis F. Parada,
        and Robert A. Weinberg, "Tumorigenic Conversion of Primary
        Embryo Fibroblasts Requires at Least Two Cooperating On-
        cogenes," *Nature* 304 (18 August 1983): 596–602; H. Earl Ruley,
        "Adenovirus Early Region 1A Enables Viral and Cellular Trans-
        forming Genes to Transform Primary Cells in Culture," ibid.,
        602–606; Robert F. Newbold and Robert W. Overell, "Fibroblast
        Immortality Is a Prerequisite for Transformation by EJ c-Ha-*ras*
        Oncogene," ibid., 648–51; "Step by Step into Carcinogenesis,"
        ibid., 582–83.

75   "Other types of . . ." Kristin White, "Chromosomes and Can-
        cer," *Medical World News* (1 March 1982): 66–87.

77   "Studies of Wilms' . . ." Yunis, "The Chromosomal Basis of
        Human Neoplasia," 229."

78   "One possible marker . . ." Joan Arehart-Treichel, "Cancer in
        the Family," *Science News* 119 (9 May 1981): 297.

79   "Melanoma, a deadly . . ." Ronald T. Acton, "The Use of Genes
        at the Major Histocompatibility Complex to Predict Risk of Ma-
        lignant Diseases and Outcome of Therapy," paper delivered at
        the American Cancer Society's 25th Science Writers' Seminar in
        San Diego, March 23, 1983.

79   "At NCI, Curtis . . ." Curtis C. Harris, "Biochemical and Molec-
        ular Epidemiology of Cancer," paper delivered at the American
        Cancer Society's 25th Science Writers' Seminar in San Diego,
        March 21, 1983.

        Chapter 9   **Limited Options**
81   ". . . if this thing . . ." C. P. Snow, "Human Care," *JAMA* 225 (6
        August 1973): 620.

82   "Available treatment approaches . . ." The treatment options dis-
        cussed in this chapter are summarized by Leon E. Rosenberg in

"Therapeutic Modalities for Genetic Diseases: An Overview," in *Progress in Clinical and Biological Research* 34, C. Papadatos and C. Bartsocas, eds. (New York: A. R. Liss, 1979), 41–79. See also Stanbury et al., *The Metabolic Basis of Inherited Disease.*

85 "A human enzyme . . ." Roscoe O. Brady et al., "Replacement Therapy for Inherited Enzyme Deficiency. Use of Purified Glucocerebrosidase in Gaucher's Disease," *New England Journal of Medicine* 291 (7 November 1974): 989–93.

86 "Similar temporary effects . . ." Nyhan, *The Heredity Factor*, 235–36.

87 "Sickle cell anemia . . ." Thomas H. Maugh II, "A New Understanding of Sickle Cell Emerges," *Science* 211 (16 January 1981): 265–67.

88 "With this understanding . . ." Thomas H. Maugh II, "Sickle Cell (II): Many Agents Near Trials," *Science* 211 (30 January 1981): 468–70.

88 "Research suggests that . . ." Joan Arehart-Treichel, "Duchenne Muscular Dystrophy: A Cure in Sight?," *Science News* 123 (15 January 1983): 42–43.

88 "In 1934 Norwegian . . ." Nyhan, *The Heredity Factor*, 218–229.

90 "In 1975 physicians . . ." "Human Fetus Is Successfully Treated," *Science News* 108 (23 August 1975): 121.

90 "In 1980 physicians . . ." "Fetus Successfully Treated in Womb," *Science News* 119 (23 May 1981): 326.

91 "These are the . . ." "New Crop of PKU Victims: Babies of Successfully Treated Girls," *Medical World News* (23 November 1981): 48.

91 "Patients with the . . ." Nyhan, *The Heredity Factor*, 236; M. Philippart et al., "Studies on the Metabolic Control of Fabry's Disease Through Kidney Transplantation," in *Sphingolipids, Sphingolipidoses and Allied Disorders*, Bruno W. Volk and Stanley M. Aronson, eds. (New York: Plenum Press, 1972), 641–49.

92 "Drugs are available . . ." William L. Nyhan, interview, July 30, 1981.

92 "The first thymic . . ." Nyhan, *The Heredity Factor*, 237–42.

93 "In 1982 physicians . . ." E. D. Thomas et al., "Marrow Transplantation for Thalassemia," *Lancet* 2 (31 July 1982): 227–29.

94 "But many groups . . ." Yair Reisner et al., "Transplantation for Severe Combined Immunodeficiency with HLA-A, B, D, DR Incompatible Parental Marrow Cells Fractionated by Soybean Agglutinin and Sheep Red Blood Cells," *Blood* 61 (February 1983): 341–48; Nancy Collins, telephone interview, May 23, 1983.

PAGE
94   "Researchers at Guy's . . ." John Newell, "Universal Trans-
        plant," *Omni* (June 1982): 42.

95   "Yale University geneticist . . ." Rosenberg, "Therapeutic Mo-
        dalities," 51.

                    Chapter 10   **Isolating Genes**

99   "Just as our . . ." Paul Berg, "Dissections and Reconstructions
        of Genes and Chromosomes," *Science* 213 (17 July 1981): 302.

101   "The isolated toad . . ." Donald D. Brown, "The Isolation of
        Genes," *Scientific American* 229 (August 1973): 20–29.

101   "In another early . . ." Gerald L. Wick, "Molecular Biology:
        Moving Toward an Understanding of Genetic Control," *Science*
        167 (9 January 1970): 157–59.

104   "Using the DNA . . ." D. J. Jolly et al., "Isolation of a Genomic
        Clone Partially Encoding Human HGPRT," *Proceedings of the
        National Academy of Sciences* 79 (August 1982): 5038–41; The-
        odore Friedmann, interview, July 28, 1982; Friedmann, "Cloning
        the HGPRT Gene," paper delivered at the Agouron Institute in
        San Diego, May 24, 1982.

105   "Since the researchers . . ." For a description of other methods
        used to retrieve genes of identifiable function after transfer into
        animal cells, see Angel Pellicer et al., "Altering Genotype and
        Phenotype by DNA-Mediated Gene Transfer," *Science* 209 (19
        September 1980): 1418–19.

106   "Initially, people got . . ." Richard Mulligan, interview, October
        19, 1982.

107   "Today at the California . . ." Leroy Hood and Michael
        Hunkapiller, "Biotechnology and Medicine of the Future," *En-
        gineering and Science*, California Institute of Technology 46
        (March 1983): 9–11; Hunkapiller, telephone interview, April 12,
        1983.

108   "With the available . . ." Leroy Hood, interview, July 29, 1981.

108   "By the end of . . ." Yvonne Baskin, "GenBank: Storehouse for
        Life's Secret Code," *Science Digest* (May 1983): 94–95.

109   "A GenBank rival . . ." Margaret Dayhoff, telephone interview,
        November 9, 1982.

110   "The Caltech team . . ." Michael Hunkapiller, telephone inter-
        view, April 12, 1983.

110   "'DNA cloning and . . ." Walter Bodmer, "The William Allan
        Memorial Award Address: Gene Clusters, Genome Organization,
        and Complex Phenotypes. When the Sequence Is Known, What
        Will It Mean?," *American Journal of Human Genetics* 33 (Sep-
        tember 1981): 675.

PAGE

111 "Many of the five . . ." Victor A. McKusick, "The Anatomy of the Human Genome," *The American Journal of Medicine* 69 (August 1980): 267–76; McKusick, telephone interview, April 11, 1983.

111 "Many more traits . . ." Frank H. Ruddle and Raju S. Kucherlapati, "Hybrid Cells and Human Genes," *Scientific American* 231 (July 1974): 36–44.

112 "Mary E. Harper at . . ." Mary E. Harper, "Direct Analysis of the Location of Human Cellular Oncogenes and Their Relationship to Chromosome Abnormalities in Tumor Cells," paper presented at the American Cancer Society's 25th Science Writers' Seminar in San Diego, March 21, 1983; Harper, interview, March 21, 1983.

112 "With techniques like . . ." McKusick, "The Anatomy of the Human Genome," 276.

112 "A group at Israel's . . ." Judy Lieman-Hurwitz et al., "Human Cytoplasmic Superoxide Dismutase cDNA Clone: A Probe for Studying the Molecular Biology of Down Syndrome," *Proceedings of the National Academy of Sciences* 79 (May 1982): 2808–11.

Chapter 11   Searching for Landmarks

114 "The next decade . . ." Mark Skolnick, "The Utah Genealogical Data Base: A Resource for Genetic Epidemiology," *Banbury Report 4–Cancer Incidence in Defined Populations* (1980): 294.

114 "University of Utah . . ." Ibid., 285–96; "Utah Genetics Research Unlocks Mysteries of Heredity," *Health Sciences Report*, University of Utah 5 (January/February 1979): 4–5.

117 "In 1978 biologists . . ." The material in this chapter on RFLP "landmark" mapping is drawn from several sources: David Botstein et al., "Construction of a Genetic Linkage Map in Man Using Restriction Fragment Length Polymorphisms," *American Journal of Human Genetics* 32 (May 1980): 314–331; Roger Lewin, "Jumping Genes Help Trace Inherited Diseases," *Science* 211 (13 February 1981): 690–92; Arlene Wyman, interview, October 22, 1982; Raymond L. White, telephone interview, April 7, 1983.

119 "In 1978 Yuet . . ." Yuet Wai Kan and Andrée Dozy, "Polymorphism of DNA Sequence Adjacent to Human Beta-Globin Structural Gene: Relationship to Sickle Mutation," *Proceedings of the National Academy of Sciences* 75 (November 1978): 5631–35; Jean L. Marx, "Restriction Enzymes: Prenatal Diagnosis of Genetic Disease," *Science* 202 (8 December 1978): 1068–69.

PAGE
119   "Then in mid-1982 . . ." Stuart H. Orkin et al., "Improved Detec-
       tion of the Sickle Mutation by DNA Analysis: Application to Pre-
       natal Diagnosis," *New England Journal of Medicine* 307 (1 July
       1982): 32–36; Judy C. Chang and Yuet Wai Kan, "A Sensitive
       New Prenatal Test for Sickle Cell Anemia," ibid., 30–32.
119   "In 1983 a team . . ." Brenda J. Conner et al., "Detection of
       Sickle Cell Beta-Globin Allele by Hybridization with Synthetic
       Oligonucleotides," *Proceedings of the National Academy of Sci-
       ences* 80 (January 1983): 278–82.
120   "In 1982 a group . . ." J. M. Murray et al., "Linkage Relationship
       of a Cloned DNA Sequence on the Short Arm of the X Chromo-
       some to Duchenne Muscular Dystrophy," *Nature* 300 (4 Novem-
       ber 1982): 69–71.
124   "Teams led by . . ." Arehart-Treichel, "Duchenne Muscular
       Dystrophy," 42–43.
125   "Mark Skolnick, Director . . ." House Committee on Science
       and Technology, *Human Genetic Engineering: Hearings Before
       the Subcommittee on Investigations and Oversight*, 97th Cong.,
       2d sess. (16–18 November 1982): 244.
125   "Medicine is becoming . . ." Ibid., 256.

       Chapter 12   **Screening the Blueprint**
126   "It is the destiny . . ." Roger L. Shinn, "Ethical Issues in Genetic
       Choices," in *Genetic Responsibility*, Mack Lipkin, Jr., and Peter
       T. Rowley, eds. (New York: Plenum Press, 1974), 116.
127   "You won't find . . ." Hood, interview, July 29, 1981; Hood and
       Hunkapiller, "Biotechnology and Medicine of the Future," 11;
       Robert S. Ledley et al., "Automated Genetic Screening," paper
       presented at the symposium on Issues Arising from the Expan-
       sion of Genetic Screening and Genetic Engineering Technology,
       American Association for the Advancement of Science Annual
       Meeting in Detroit, May 29, 1983.
127   "Since 1967, when . . ." For a summary of legal decisions, see
       National Institute of Child Health and Human Development, *An-
       tenatal Diagnosis*, I-165-71.
128   "By 1981 some . . ." Sharon R. Stephenson and David D.
       Weaver, "Prenatal Diagnosis—A Compilation of Diagnosed
       Conditions," *American Journal of Obstetrics and Gynecology*
       141 (1 October 1981): 319–36.
128   "Since the early . . ." Neil A. Holtzman, *Newborn Screening for
       Genetic-Metabolic Diseases*, DHEW Publication No. (HSA) 77–
       5207 (Washington: GPO, 1977), 17.
128   "Virtually all states . . ." Department of Health and Human Ser-
       vices, *State Laws and Regulations on Genetic Disorders*, DHHS
       Publication No. (HSA) 81-5243 (Washington: GPO, 1980).

PAGE

129 "The nation got . . ." Nyhan, *The Heredity Factor*, 86–8.

130 "A German researcher . . ." National Institute of Child Health and Human Development, *Antenatal Diagnosis*, I-55-7.

131 "Machines for fully . . ." Ibid., I-78.

132 "When it was . . ." Ibid., I-114.

132 "Another new approach . . ." Gina Bari Kolata, "First Trimester Prenatal Diagnosis," *Science* 221 (9 September 1983): 1031–32.

133 "In the first . . ." National Institute of Child Health and Human Development, *Antenatal Diagnosis*, I-4 and 59.

133 "In a report issued . . ." Ibid., I-15-6 and I-201-3.

134 "In 95 percent . . ." Ibid., I-71-3.

135 "The 1979 report . . ." Ibid., I-80-7.

135 " 'It will undoubtedly . . .' " Ibid., I-221.

136 "Unlike many of . . ." Ibid., I-143; Nyhan, *The Heredity Factor*, 174–77.

137 "Medical geneticist Michael . . ." Julie Ann Miller, "Update on Tay-Sachs Screening," *Science News* 120 (31 October 1981): 282.

137 "This disease is . . ." National Institute of Child Health and Human Development, *Antenatal Diagnosis*, I-157.

138 "The congressional Office . . ." Constance Holden, "Looking at Genes in the Workplace," *Science* 217 (23 July 1982): 336–37.

138 "Subcommittee Chairman Representative . . ." Ruth Marcus, *The Washington Post* (23 June 1982).

Chapter 13 The Genetic Shuffle

143 "Once we thought . . ." Maxine Singer, "Recombinant DNA Revisited," *Science* 209 (19 September 1980): 1317.

145 "Some researchers have . . ." Jean L. Marx, "Tracking Genes in Developing Mice," *Science* 215 (1 January 1982): 44–47.

146 "Another spin-off of . . ." Yvonne Baskin, "In Search of the Magic Bullet," *Technology Review*, Massachusetts Institute of Technology 85 (October 1982): 18–22.

147 "Plant genetic engineers . . ." Leslie Roberts, "Gardening in Test Tubes," *Mosaic*, National Science Foundation 13 (May/June 1982): 14–15.

147 "One approach is . . ." For more information on the use of liposomes, see Robert Fraley and Demetrios Papahadjopoulos, "New Generation Liposomes: The Engineering of an Efficient Vehicle for Intracellular Delivery of Nucleic Acids," *Trends in Biochemical Sciences* (March 1981): 77–80.

147 "Another technique is . . ." Rozanne M. Sandri-Goldin et al., "High Frequency Transfer of Cloned Herpes Simplex Virus Type I Sequences to Mammalian Cells by Protoplast Fusion," *Molecular and Cellular Biology* 1 (August 1981): 743–52.

PAGE   *Chapter 14* **The Friendly Virus**

149   "We live in . . ." Lewis Thomas, *The Lives of a Cell* (New York: Bantam Books, 1975), 4.

150   "Viruses that infect . . ." Allan M. Campbell, "How Viruses Insert Their DNA into the DNA of the Host Cell," *Scientific American* 235 (December 1976): 102–13.

150   "What Lederberg discovered . . ." M. L. Morse, Esther M. Lederberg, and Joshua Lederberg, "Transductional Heterogenotes in *Escherichia coli*," *Genetics* 41 (1956): 758–59.

151   "The possibilities for . . ." Lederberg, "Participatory Evolution," 50.

151   "'One can imagine . . ." Rollin D. Hotchkiss, "Portents for a Genetic Engineering," *The Journal of Heredity* 56 (September–October 1965): 199.

151   "The saga of . . ." The material in this section was compiled from a number of sources: Stanfield Rogers, telephone interview, October 1, 1981; Rogers, "Reflections on Issues Posed by Recombinant DNA Molecule Technology. II," *Annals of the New York Academy of Sciences* 265 (23 January 1976): 66–70; Rogers, "Gene Therapy for Human Genetic Disease?," *Science* 178 (10 November 1972): 648–49; Theodore Friedmann and Richard Roblin, "Gene Therapy for Human Genetic Disease?," *Science* 175 (3 March 1972): 949–55; Rogers, "Genetic Engineering," *Human Genetics*, Proceedings of the 4th International Congress of Human Genetics, Paris (6–11 September 1971): 36–40; Rogers, "Gene Therapy: A Potentially Invaluable Aid to Medicine and Mankind," *Research Communications in Chemical Pathology and Pharmacology* 2 (July–September 1971): 587–600; Rogers, "Skills for Genetic Engineers," *New Scientist* (29 January 1970): 194–96; Rogers, "Shope Papilloma Virus: A Passenger in Man and Its Significance to the Potential Control of the Host Genome," *Nature* 212 (10 December 1966): 1220–22.

152   "A report of . . ." "Good Genes for Bad," *Newsweek* 69 (19 June 1967): 92.

152   "'It was clear . . ." Rogers, "Reflections on Issues," 66.

152   "'How we wished . . ." Rogers, "Genetic Engineering," 37.

153   "'The older child . . ." Rogers, telephone interview, October 1, 1981.

154   "'The dose of . . ." Rogers, "Reflections on Issues," 67.

154   "'A short time . . ." Ibid.

154   "'Although these results . . ." Ibid.

155   "In 1971 Carl . . ." Carl R. Merril, Mark R. Geier, and John C. Petricciani, "Bacterial Virus Gene Expression in Human Cells," *Nature* 233 (8 October 1971): 398–400.

PAGE   *Chapter 15* **The Monkey Virus**

156   "Work on recombinant . . ." François Jacob, *The Possible and the Actual* (New York: Pantheon, 1982), 46.

156   "Stanford University biochemist . . ." Material for this chapter was drawn from several sources: Mulligan, interview, October 19, 1982; Berg, "Dissections and Reconstructions of Genes and Chromosomes," 296–303; Mulligan and Berg, "Expression of a Bacterial Gene in Mammalian Cells," *Science* 209 (19 September 1980): 1422–27.

158   "'What we were . . .'" Mulligan, interview, October 19, 1982.

158   "The answer was . . ." Jean L. Marx, "Successful Transplant of a Functioning Mammalian Gene," *Science* 202 (10 November 1978): 610.

   Chapter 16 **Soaking Up Naked DNA**

160   "One of the lessons . . ." P. B. Medawar, "Ethical Considerations, Discussion," in *Man and His Future*.

160   "Leaning on the . . ." "Heredity by Injection," *Time* 69 (10 June 1957): 81.

162   "In 1944 he published . . ." Oswald T. Avery, Colin M. Mac-Leod, and Maclyn McCarty, "Studies on the Chemical Nature of the Substance Inducing Transformation of Pneumococcal Types," *Journal of Experimental Medicine* 79 (1 February 1944): 137–57.

162   "'After we had . . .'" Hotchkiss, "Portents for a Genetic Engineering," 198.

163   "The prospects for . . ." Jean L. Marx, "Gene Transfer in Mammalian Cells: Mediated by Chromosomes," *Science* 197 (8 July 1977): 146–48.

164   "In 1977 a group . . ." Richard Axel, interview, October 13, 1982; Angel Pellicer et al., "Altering Genotype and Phenotype by DNA-Mediated Gene Transfer," *Science* 209 (19 September 1980): 1414–22.

165   "Next, Berg and . . ." Mulligan, interview, October 19, 1982; Mulligan and Berg, "Expression of a Bacterial Gene in Mammalian Cells."

166   "Berg quickly quashed . . ." Jean L. Marx, "Gene Transfer Moves Ahead," *Science* 210 (19 December 1980): 1336.

   Chapter 17 **New Genes for Mice and Men**

168   "Paradoxical forces exist . . ." Arno G. Motulsky, "Brave New World," *Science* 185 (23 August 1974): 654.

PAGE

169   "Using the transformation . . ." Jean L. Marx, "Gene Transfer Given a New Twist," *Science* 208 (25 April 1980): 386–87; Martin J. Cline et al., "Gene Transfer in Intact Animals," *Nature* 284 (3 April 1980): 422–25; Karen E. Mercola et al., "Insertion of a New Gene of Viral Origin into Bone Marrow Cells of Mice," *Science* 208 (30 May 1980): 1033–35.

170   "The applications to . . ." Marx, "Gene Transfer Given a New Twist," 387.

170   "And it was 'theoretically . . ." Mercola et al., "Insertion of a New Gene of Viral Origin into Bone Marrow Cells of Mice," 1035.

170   "For almost a year . . ." The account of Cline's human experiments was drawn from several sources: Martin Cline, interview, July 28, 1981; National Institutes of Health Memorandum, from Executive Secretary, NIH Ad-Hoc Committee on the UCLA Report Concerning Certain Research Activities of Dr. Martin J. Cline, to The Record, 21 May 1981; Nicholas Wade, "Gene Therapy Caught in More Entanglements," *Science* 212 (3 April 1981): 24–25; Wade, "UCLA Gene Therapy Racked by Friendly Fire," *Science* 210 (31 October 1980): 509–11; Gina Bari Kolata and Wade, "Human Gene Treatment Stirs New Debate," *Science* 210 (24 October 1980): 407; W. French Anderson and John C. Fletcher, "Gene Therapy in Human Beings: When Is It Ethical to Begin?," *New England Journal of Medicine* 303 (27 November 1980): 1293–97; Karen E. Mercola and Cline, "The Potentials of Inserting New Genetic Information," *New England Journal of Medicine* 303 (27 November 1980): 1297–1300.

173   "'Human Engineering: Pioneer . . ." Paul Jacobs, "Human Engineering: Pioneer Genetic Implants Revealed," *Los Angeles Times* (8 October 1980).

173   "Genetic engineering has . . ." Associated Press (9 October 1980).

173   "'The experiment was . . ." James Gorman, "Bad Blood Over Good Genes," *Discover* (December 1980): 81.

173   "'There is simply . . ." Wade, "UCLA Gene Therapy Racked by Friendly Fire," 509.

174   "'I'm a doctor . . ." Gorman, "Bad Blood Over Good Genes," 81.

174   "'When do you . . ." Jacobs, "Human Engineering: Pioneer Genetic Implants Revealed."

174   "Fellow clinician Theodore . . ." House Committee on Science and Technology, *Human Genetic Engineering*, 296–97.

175   "Reflecting on the . . ." Cline, interview, July 28, 1981.

PAGE
175   "It was a formality . . ." National Institutes of Health Memorandum, 16.

175   "As a result, he . . ." Nicholas Wade, "Gene Therapy Pioneer Draws Mikadoesque Rap," *Science* 212 (12 June 1981): 1253.

175   "The NIH appeals . . ." Marjorie Sun, "Martin Cline Loses Appeal on NIH Grant," *Science* 218 (1 October 1982): 37.

175   " 'If it works . . .' " Cline, interview, July 28, 1981.

175   "When Cline appeared . . ." House Committee on Science and Technology, *Human Genetic Engineering*, 449–51.

Chapter 18 Injecting Embryos

177   "Each new power . . ." C. S. Lewis, *The Abolition of Man* (New York: Macmillan, 1965), 71.

177   "One scientist who . . ." Jon Gordon, interview, October 14, 1982.

178   "A British biologist . . ." John B. Gurdon, "Transplanted Nuclei and Cell Differentiation," *Scientific American* 219 (December 1968): 24–35.

178   "Karl Illmensee of . . ." Karl Illmensee and Peter C. Hoppe, "Nuclear Transplantation in *Mus musculus*," *Cell* 23 (January 1981): 9–18; Jean L. Marx, "Three Mice 'Cloned' in Switzerland," *Science* 211 (23 January 1981): 375–76.

179   "In 1983, however . . ." Jean L. Marx, "Swiss Research Questioned," *Science* 220 (3 June 1983): 1023.

179   "While the charges . . ." James McGrath and Davor Solter, "Nuclear Transplantation in the Mouse Embryo by Microsurgery and Cell Fusion," *Science* 220 (17 June 1983): 1300–1302.

179   "Others injected copies . . ." Charles Lane, "Rabbit Hemoglobin from Frog Eggs," *Scientific American* 235 (August 1976): 60–69.

180   "In 1979 and 1980 . . ." Marx, "Gene Transfer Moves Ahead," 1335; W. French Anderson and Elaine G. Diacumakos, "Genetic Engineering in Mammalian Cells," *Scientific American* 245 (July 1981): 106–21.

180   "Since 1974 Rudolf . . ." Brigid Hogan and Jeffrey Williams, "Integration of Foreign Genes into the Mammalian Germ Line: Genetic Engineering Enters a New Era," *Nature* 294 (5 November 1981): 9–10.

181   "In 1980 Jon . . ." Jon Gordon et al., "Genetic Transformation of Mouse Embryos by Microinjection of Purified DNA," *Proceedings of the National Academy of Sciences* 77 (December 1980): 7380–84; Gordon, interview, October 14, 1982.

181   "In 1981 at least . . ." Jean L. Marx, "Still More About Gene Transfer," *Science* 218 (29 October 1982): 459–60; Marx, "More Progress on Gene Transfer," *Science* 213 (28 August 1981): 996–97.

PAGE
181   "A flurry of headlines . . ." "Gene Transferred Between Spe-
       cies," *Los Angeles Times* (8 September 1981); Jean L. Marx,
       "Globin Gene Transferred," *Science* 213 (25 September 1981):
       1488.
183   "We don't know . . ." Gordon, interview, October 14, 1982.

       Chapter 19   **Building Mighty Mice**
185   "Since we surely . . ." Daniel Callahan, "Ethical Responsibility
       in Science in the Face of Uncertain Consequences," *Annals of
       the New York Academy of Sciences* 265 (1976): 1–12.
185   "An animal responds . . ." This chapter was compiled from sev-
       eral sources: Richard Palmiter, interview, November 30, 1982;
       Palmiter et al., "Dramatic Growth of Mice That Develop from
       Eggs Microinjected with Metallothionein-Growth Hormone Fu-
       sion Genes," *Nature* 300 (16 December 1982): 64–65; Jean L.
       Marx, "Building Bigger Mice Through Gene Transfer," *Science*
       218 (24 December 1982): 1298.
188   "(Although, as geneticist . . ." House Committee on Science and
       Technology, *Human Genetic Engineering*, 464–65.

       Chapter 20   **The Genetic Bureaucracy**
190   "What distinguishes a . . ." Jacob, *The Possible and the Actual*,
       41.
190   "Only a 1 percent . . ." Ibid., 42.
194   "The next question . . ." The section on gene regulation in higher
       animals was drawn from a number of sources: Axel, interview,
       October 13, 1982; Thomas Maniatis, interview, October 18, 1982;
       William Check, "The Regulation of Gene Expression," *Mosaic*,
       National Science Foundation 13 (November/December 1982):
       29–34; Gina Bari Kolata, "Genes Regulated Through Chromatin
       Structure," *Science* 214 (13 November 1981): 775–76; Jean L.
       Marx, "Gene Control Puzzle Begins to Yield," *Science* 212 (8
       May 1981): 653–55; Roger Lewin, "How Conversational Are
       Genes?," *Science* 212 (17 April 1981): 313–15; Lewin, "Biggest
       Challenge Since the Double Helix," *Science* 212 (3 April 1981):
       28–32.
198   "'Something is still . . ." Axel, interview, October 13, 1982.
198   "Or, Maniatis speculates . . ." Maniatis, interview, October 18,
       1982.
198   "In 1983 Japanese . . ." Hisato Kondoh, Kunio Yasuda, and T.
       S. Okada, "Tissue-Specific Expression of a Cloned Chick Delta-
       Crystallin Gene in Mouse Cells," *Nature* 301 (3 February 1983):
       440–42.

PAGE *Chapter 21* **Red-Eyed Flies**

201 "It doesn't take . . ." Information for the chapter on gene transfer in fly embryos was taken from Allan C. Spradling and Gerald M. Rubin, "Transposition of Cloned P Elements into *Drosophila* Germ Line Chromosomes," *Science* 218 (22 October 1982): 341–47; Rubin and Spradling, "Genetic Transformation of *Drosophila* with Transposable Element Vectors," ibid., 348–53; and Jean L. Marx, "Gene Transfer into the *Drosophila* Germ Line," ibid., 364–65.

Chapter 22 **Taming the Retrovirus**

204 "As in all human . . ." Hermann J. Muller, "The Prospects of Genetic Change," *American Scientist* 47 (December 1959): 561.

205 "'The infected cell . . ." Most of the material in this chapter came from Mulligan, interview, October 19, 1982, and telephone interviews with Mulligan, April 27 and August 25, 1983.

205 "The chromosomes of . . ." Raoul E. Benveniste and George J. Todaro summarize and list references on gene transfers by viruses in a letter, "Gene Transfer Between Eukaryotes," *Science* 217 (24 September 1982): 1202.

Chapter 23 **Healing with Genes**

215 "Not only the . . ." Hotchkiss, "Portents for a Genetic Engineering," 197.

217 "In 1980 physician . . ." Anderson and Fletcher, "Gene Therapy in Human Beings: When Is It Ethical to Begin?," 1295.

220 "The process is . . ." John H. Richards, "Structure and Function in Biochemistry," *Engineering and Science*, California Institute of Technology 46 (March 1983): 15–17.

220 "In April 1982 . . ." Yuet Wai Kan, interview, September 10, 1982; Gary F. Temple et al., "Construction of a Functional Human Suppressor tRNA Gene: An Approach to Gene Therapy for Beta-Thalassemia," *Nature* 296 (8 April 1982): 537–40.

221 "Another research team . . ." Phillip A. Sharp, interview, October 20, 1982; Robert M. Hudziak et al., "Establishment of Mammalian Cell Lines Containing Multiple Nonsense Mutations and Functional Suppressor tRNA Genes," *Cell* 31 (November 1982): 137–46.

223 "'There are fifty . . ." Gordon, interview, October 14, 1982.

224 "Bernard D. Davis of . . ." Bernard D. Davis, "The Two Faces of Genetic Engineering in Man," *Science* 219 (25 March 1983): 1381.

PAGE
225   "During that year . . ." The section on fetal gene reactivation was taken from Timothy J. Ley et al., "5-Azacytidine Selectively Increases Gamma-Globin Synthesis in a Patient with Beta+ Thalassemia," *New England Journal of Medicine* 307 (9 December 1982): 1469–75; Edward Benz, Jr., "Clinical Management of Gene Expression," ibid., 1515–16; and Gina Bari Kolata, "Fetal Hemoglobin Genes Turned on in Adults," *Science* 218 (24 December 1982): 1295–96.

225   " 'The ability to . . .' " Rosenberg, "Therapeutic Modalities for Genetic Diseases," 43.

226   "(In 1965 Rollin . . ." Hotchkiss, "Portents for a Genetic Engineering," 201.

227   "Already genes for . . ." Marx, "Cloning the Genes of the MHC," 401.

228   " 'With a more . . .' " Tatum, "A Case History," 1714.

Chapter 24   Genetic Twilight?

229   "We cannot expect . . ." Joshua Lederberg, "Haldane's Biology and Social Insight," in *Haldane and Modern Biology*, K. R. Dronamraju, ed. (Baltimore: Johns Hopkins Press, 1968), 224.

229   "Isolated in a remote . . ." Information on Pimas, mice, and diabetes was compiled from several sources: James V. Neel, "The Thrifty Genotype Revisited," in *The Genetics of Diabetes Mellitus*, 283–93; Chris Land, "Probing the Plight of the Pimas," *Health Science Spectrum*,, The University of Texas Health Science Center at Dallas (Summer 1979): 10–15; Douglas L. Coleman, "Obesity Genes: Beneficial Effects in Heterozygous Mice," *Science* 203 (16 February 1979): 663–65.

230   "Since 1949 a . . ." Thomas H. Maugh II, "Why Does Sickle Trait Persist?," *Science* 211 (16 January 1981): 266.

232   " '. . . Society now comes . . .' " Hermann J. Muller, "Genetic Progress by Voluntarily Conducted Germinal Choice," in *Man and His Future*, 252.

233   "Huxley agreed with . . ." Julian Huxley, "The Future of Man— Evolutionary Aspects," in *Man and His Future*, 17.

233   " '. . . In the distant . . .' " Arno G. Motulsky, "Impact of Genetic Manipulation on Society and Medicine," *Science* 219 (14 January 1983): 136.

235   " 'Many bad mutations . . .' " Dobzhansky, *Mankind Evolving*, 139.

235   "Two decades later . . ." House Committee on Science and Technology, *Human Genetic Engineering*, 388–89.

Chapter 25   Improving the Species?

236   ". . . certainly man is . . ." Jean Rostand, "Can Man Be Modified?," *The Saturday Evening Post* (2 May 1959): 97.

PAGE

236 "Men would indeed . . ." Hermann J. Muller, "Prospects of Genetic Change," 561.

237 "Graham, then seventy-four . . ." Matt Potter, "Sperm Bank Propels Quiet Man to Fame," *San Diego Tribune* (3 March 1980).

237 "Graham had discussed . . ." William J. Broad, "A Bank for Nobel Sperm," *Science* 207 (21 March 1980): 1326–27.

237 "By the summer . . ." Paul Smith, research officer and sperm donor recruiter, telephone interview, May 4, 1983.

237 "J.B.S. Haldane noted . . ." Lederberg, "Haldane's Biology and Social Insight," 221.

238 "Prospects for improving . . ." Robert P. Laurence, "Second 'Nobel Sperm Bank' Mom Single," *San Diego Union* (16 July 1982); Laurence, "Baby Won't Face Pressure," ibid. (17 July 1982).

238 "('There is no . . ." Broad, "A Bank for Nobel Sperm," 1327.

239 "As geneticist Theodosius . . ." Dobzhansky, *Mankind Evolving*, 22.

239 " 'The closer the . . ." Ibid., 247–48.

240 "In 1963 Haldane . . ." Haldane "Biological Possibilities for the Human Species," 342–57.

241 "Geneticist Sheldon C. . . ." Sheldon C. Reed, "Eugenics Tomorrow," in *Haldane and Modern Biology*, 242.

241 "Joshua Lederberg and . . ." R. Michael Davidson, "Man's Participatory Evolution," *Current* (March 1969): 8–9.

241 "In 1959 French . . ." Rostand, "Can Man Be Modified?," 97–98.

241 "In 1982 University . . ." Easton's vision was described in Mishara, "Small People"; Dyson's by Irving Lieberman in "Freeman Dyson's Future Man," *Omni* (June 1982): 112; and organic computers by Susan Chace in "Silicon's Successor? Tomorrow's Computer May Reproduce Itself, Some Visionaries Think," *The Wall Street Journal* (6 January 1982).

242 "In 1963, listening . . ." Joshua Lederberg's comments are part of "Discussion," in *Man and His Future*, 235.

243 "In 1965 Nobel . . ." Max Perutz, "Molecular Biology of the Future," *New Scientist* 85 (31 January 1980): 298.

244 "At Columbia University . . ." Eric R. Kandel and James H. Schwartz, "Molecular Biology of Learning: Modulation of Transmitter Release," *Science* 218 (29 October 1982): 433–43; Kandel, "Small Systems of Neurons," *Scientific American* 241 (September 1979): 67–76.

245 "Moving from learning . . ." The section on behavioral genetics in Aplysia was compiled from several sources: Axel, interview,

October 13, 1982; Axel, "Genes Modulating Behavior," seminar at University of California at San Diego Medical Center, May 18, 1983; Roger Lewin, "Gene Family Controls a Snail's Egg Laying," *Science* 216 (14 May 1982): 720–21.

246   "Some 1,500 strains . . ." Dennis Meredith, "A Second Golden Age in *Drosophila* Research," *Engineering and Science*, California Institute of Technology 45 (September 1981): 13–18.

247   "In the 1960s Peter . . ." Carol Grant Gould, "Out of the Mouths of Beasts," *Science* 83 (April 1983): 69–72.

248   " 'Replacing a faulty . . ." House Committee on Science and Technology, *Human Genetic Engineering*, 287–88.

248   " 'The most important . . ." Ibid., 465–66.

249   " 'If you think . . ." Axel, interview, October 13, 1982.

249   " 'It's extremely likely . . ." Jeffrey Saver, "Edward O. Wilson: Father of a New Science," *Science Digest* (May 1982): 86.

250   "Philosopher Stephen Stich . . ." House Committee on Science and Technology, *Human Genetic Engineering*, 536–37.

251   " 'The current tendency . . ." President's Commission, *Splicing Life*, 68.

252   "Alexander Capron, executive . . ." House Committee on Science and Technology, *Human Genetic Engineering*, 187.

Chapter 26   Choosing Our Future

253   "Man and man . . ." Dobzhansky, *Mankind Evolving*, 346–47.

255   " 'Human genetic engineering . . ." House Committee on Science and Technology, *Human Genetic Engineering*, 2.

255   " 'History has shown . . ." President's Commission, *Splicing Life*, 95–96.

255   " 'Some people have . . ." Ibid., covering letter to the President.

255   "The commission also . . ." Ibid., 77.

256   " 'It is true . . ." Ibid., 2.

256   "The report concluded . . ." Ibid., 3 and covering letter.

256   " 'Gene splicing is . . ." Ibid., 21.

257   " 'In the view . . ." Ibid., 53–54.

257   "June Goodfield expressed . . ." June Goodfield, *Playing God* (New York: Random House, 1977), 177.

257   "Molecular biologist Robert . . ." Robert L. Sinsheimer, "Genetic Engineering: Life as a Plaything," *Technology Review*, Massachusetts Institute of Technology (April 1983): 14.

257   "If 'playing God' . . ." President's Commission, *Splicing Life*, 57–59.

258   "As Leon R. . . . " Leon R. Kass, "The New Biology: What Price Relieving Man's Estate?," *Science* 174 (19 November 1971): 783.

PAGE

258 "A more realistic . . ." President's Commission, *Splicing Life*, 72.

259 " 'If the promise . . ." LeRoy Walters, "Ethical Issues in Genetic and Reproductive Engineering," paper presented at the Symposium on Human Embryo and Gene Manipulation, American Association for the Advancement of Science Annual Meeting in Washington, D.C., January 4, 1982.

260 "Jon Gordon of . . ." Gordon, interview, October 14, 1982.

260 "Interest in establishing . . ." Colin Norman, "Clerics Urge Ban on Altering Germline Cells," *Science* 220 (24 June 1983): 1360–61.

261 " 'I do not . . ." John Fletcher, "Genetics, Choice and Society," in *Genetic Responsibility*, 99.

Index

abortion, fetal:
  as choice, 13, 95, 134, 137, 260; *see also*
    counseling, genetic
  and sexual choice for child, 131
  spontaneous, 38–39, 133
Acton, Ronald T., 79
adrenogenital syndrome, 84
AFP (alpha fetoprotein), 130–31
aging process, human, 53
AIDS (acquired immune deficiency
    syndrome), 71
alcoholism, 64–65
alkaptonuria, 29–30, 33
Alu repeat sequence, 105
amniocentesis, 11–12, 13, 130–35 *passim*
  for sex selection, discouraged, 131
amplification, 71
Anderson, W. French, 217, 248
anemia:
  Cooley's, *see* beta-thalassemia
  Fanconi's, 78
  favism, 90
  hemolytic, 90, 138
  juvenile pernicious, 83–84
  sickle cell, *see* sickle cell anemia
animal cells:
  fusion, 145–48
  human genes transferred to, 104–105
  hybridization, 111–12, 182
  see also microinjection
animals:
  traits of, genetically transferred to humans,
    240–41
  see also livestock
ankylosing spondylitis, 49–50, 139
antibodies:
  and cancer genes, 75
  and HLA markers, 50
  "jumping genes" of, 73–74
  monoclonal, 146
antigens, and HLA markers, 50
Aplysia (sea slug), 243–46
arthritis: ankylosing spondylitis, 49–50, 139
Asberg, Marie, 62
A-T (ataxia-telangiactasia), 78
atherosclerosis, 56–58
autoimmune diseases, 51–52, 227
autoradiograph ("fingerprint"), genetic, 121,
    126–27
availability of medical treatment:
  of gene therapy, 259

  for infectious diseases, 15
Avery, Oswald T., 162
Axel, Richard, 164, 173, 197–98, 245–46, 249

bacteria:
  as "factories" for hormones, etc., 22, 194
  gene regulation in, 193–95
  gene swapping by, 15, 143
  pirated substances from, 102
Baltimore, David, 69–70
Banting, Frederick G., 81–82
Barbacid, Mariano, 69, 72
Beadle, George, 33–34
behavior, genetic influences on, 242–50
Benoît, Jacques, 160, 161
Benz, Edward, Jr., 225
benzopyrene, as carcinogen, 79
Berg, Paul, 99, 106, 156–59, 165–66
Best, Charles H., 82
beta-thalassemia:
  carrier-screening program for, 137
  drug therapy for, 225, 226–27
  gene therapy for, 171–73, 220–21
  marrow transplantation for, 94
biotin, 90
Bishop, Michael, 68
bladder cancer, 71–72, 74
blood disorders, 170–71
  inherited, 38
  see also anemia; sickle cell anemia;
    thalassemias
Bodmer, Walter, 110–11
bone marrow transplants, 83, 93–94, 210–11
Botstein, David, 117, 120
Boyer, Herbert, 157
breast cancer, 77, 78–79
Brinster, Ralph, 185–89
Burkitt's lymphoma, 71, 73, 112

Callahan, Daniel, 185, 238
cancer, 66–80
  isolating genes of, 105
  predisposition to, 76–77, 120
  tests for responses to carcinogens, 79–80
  treatment, 146, 168–75
  see also specific forms of cancer
cancer genes, *see* oncogenes
Capecchi, Mario R., 222
Capron, Alexander, 252
carcinogens:
  activating of, to initiate cancer, 79–80